Seeing the Animal Whole

And Why It Matters

Seeing the Animal Whole

And Why It Matters

Craig Holdrege

Lindisfarne Books / 2021

Lindisfarne Books
402 Union Street No. 58, Hudson, NY 12534
an imprint of SteinerBooks / www.steinerbooks.org

All drawings and photographs are by Craig Holdrege unless otherwise noted.
Cover design by Mary Giddens and Craig Holdrege.
Cover photo: A three-toed sloth "flowing" up a tree in Brazil.

Print ISBN: 978-1-58420-903-4
Ebook ISBN: 978-1-58420-904-1

Printed in the United States of America

FOR MY GRANDCHILDREN

Sebastian, Ava, Anika, and Jonah

Contents

Introduction

LATE ONE AFTERNOON IN JUNE, my wife and I were at an overlook in the northern part of Yellowstone National Park. We peered down into the gorge of the Yellowstone River, which was full of rushing, muddy spring meltwater. The gorge is capped by a nearly vertical wall of basalt, and below it a steep and gradually eroding incline of ancient lava rises above the riverbed. Some trees grow on these steep canyon walls, but very little herbaceous vegetation.

After a while, we noticed that there were animals on those cliff sides. They were female bighorn sheep (ewes) and their young lambs. Do not picture domestic sheep. The ewes were sleek and fairly short-haired, and resemble domestic goats more than sheep. The lambs were grey and had long spindly legs. In coloration the sheep blended well with their rocky ground. Moving slowly among the rock-strewn walls—walls on which, as a human being, I would certainly want to be roped in and secured— the ewes fed occasionally on the sparse vegetation. They often paid little attention to their lambs, which were frolicking around on the inclines, giving no attention to the deadly chasm into which they could plunge. The lambs jumped from rock to rock, ran after one another, and some- times kicked up rocks that bounded down into the river far below.

These bighorn sheep inhabit a remarkably inhospitable environment —and yet, it is their home, their place. Inhospitable means inhospitable for us human beings. It was so vivid that they live in a different world, not only in terms of the outer environment, but also in the way they are part of that environment. They are not worrying about falling; the ewes are not agonizing over the wild behavior of their lambs; they are not con- cerned at how sparse the vegetation is; they are not planning ahead to decide where to spend the night. They live embedded in this "harsh" environment as an extension of their being. They are at home.

On this trip, while observing so many different animals, such different ways of being and ways of relating to the surroundings, I had the growing sense that I hardly fathom what it is to be an animal—to be so integrally entwined with one's world. We have kinship as living, sentient beings, but I'm the one observing them and thinking about their characteristics. I can appreciate their existence, and, in fact, I'm in awe of them. This connects me with them and at the same time makes me realize how different I am.

I have been teaching about animals and studying them for many years. It's interesting how growing knowledge has increased my wonder at every meeting with an animal — that is, at every meeting when I am really attentive and get out of myself enough to realize what a fascinating and in a sense unfathomable creature I'm attending to.

I would not have written this book unless I felt I had something worthwhile to say about the nine animals I portray. That's obvious. One of my intellectual ancestors, who continues to inspire my work, is the poet and scientist Johann Wolfgang von Goethe. You'll be hearing more from him. In the 1780s, while on a two-year journey in Italy, he wrote to a friend, "After what I have seen of plants and fish in Naples and Sicily, I would be tempted—were I ten years younger—to undertake a journey to India, not to discover something new, but to view in my way what has been discovered."[1] For me the study of animals has been in this vein. It's about a different way of looking at and thinking about animals—and life in general—that I believe opens up fresh insights and a heightened respect for our fellow creatures.

I do not plan to talk about the approach in an abstract way. It should become clear in the course of the book, and I have written about it in depth elsewhere.[2] But let me just say a few things upfront. As a child I felt connected with the natural world in the sense that it was the background and basis of my play. I grew up in Colorado, and I breathed in the mountain air, avoided the cacti in foothills, and loved making forts in the rocks. But I wasn't a born naturalist who collected and observed bugs or birds. By my teenage years I was pretty well alienated from everything, although being in nature was always something special. But I didn't know where and how to find meaning. My introduction to biology was through dissecting a frog, an activity that by no means drew me to study science.

My question—at the time more felt than articulated—was: Is it possible to connect with the world in a clear and authentic way that does justice to what I'm engaging with? The exploitation of humans and the rest of nature weighed, and still weighs, heavily on me.

I turned to the study of philosophy and saw more clearly that much of the problem is how we view the world—what kinds of stories we create about the nature of reality and the place of humanity in the larger scheme of things. I encountered myriad worldviews, and in learning to think more critically about how we think about things, I realized that modern science is also a worldview. It does not talk about the world "as it is" but the world through an array of assumptions that we need to understand.

While much philosophy was illuminating, it also left me feeling one-sided. How do I get to the world in which we actually live and which supports our existence? Luckily I encountered the writings of Owen Barfield and Rudolf Steiner's early philosophical works based on his study of Goethe as a scientist.[3] Here I found an approach that was rigorous and at the same time freeing. I could see how it is possible to discover ever more meaning in the world around me.

Goethe lived from 1749 to 1832 and is most well known for his literary writings. But he was also the founder of what we can call today a holistic and phenomenological approach to science. In recent decades his method of "delicate empiricism" has become more widely valued and discussed.[4] Goethe had an ecological view of the world in the sense that he realized that to understand any given phenomenon, you need to consider the larger web of relations in which it is embedded. And he saw science as a way in which the human being participates in nature.

Over time, the larger field of holistic and phenomenological science opened up to me and continues to fructify my work.[5] I went to Europe and studied with and learned from a number of Goethean-inspired biologists and educators.[6] I have worked with this approach in my own way for many decades now. What has become ever clearer to me is that we can practice science as a way of participating in the world. It is not about becoming a dispassionate onlooker. Yes, we need rigor and the ongoing work to bring the ideas that grow in us into ever-closer relation to the things themselves. But instead of coming up with elaborate speculative theories that we place between us and the richness of the world, we can school our imagination and thinking to become concrete, to adapt itself to what we perceive, and thereby gain insights into connections that are at first not evident to us.

After my studies, I became a high school teacher, and over the course of 21 years teaching zoology I was always searching for ways to bring animals closer to my students. Along the way I became fascinated with some of the animals that I will portray here.

Only later, when I helped cofound The Nature Institute in upstate New York, where I have worked ever since, did I have the opportunity for more in-depth research and for travels to experience the animals in the wild. I have written articles, book chapters, and short monographs on which many chapters of this book are based.[7] At the same time, I have been developing the Goethean approach and working with it in courses with university students, educators and other professionals. This book brings the fruits of these efforts.

Each of the nine animals I portray in the following chapters has its own unique character, whether the slow sloth, the burrowing mole, the towering giraffe, or the huge but flexible elephant. When an animal attracts our interest, we may experience its beauty or perhaps its strangeness and otherness. But in any case the animal is a unity, a presence and force in the world. It is one thing to feel all this, it is another to comprehend more clearly the nature of these marvelous creatures and bring their character to expression. That's what I want to do in this book.

My aim is to give voice to these animals as beings rich in qualities that make them distinctive and irreplaceable. Each of the portrayals is self-contained and illuminates the specific way of being of that animal. You will encounter many fascinating details about these animals. But I don't want to bombard you with collections of facts. I want to show how an animal's many features are interconnected and are a revelation of the animal as a whole. Moreover, you will see how the animal is intimately interwoven with the larger context that supports its life, a context that it also actively influences.

We do great injustice to animals when we make them into embodiments of our theories (as when we look at them as evolutionary survival strategies) or project our all-too-human characteristics onto them. With the open-ended question "Who are you?" and the will to let the animals themselves be my guide, I avoid the mechanistic and anthropomorphic interpretations that unfortunately take hold of so much writing about animals.

I have increasingly come to see that animals are beings that actively orchestrate their existence. I don't mean only in the flexibility of their responses and behavior, which is quite obvious. I mean that we can discover dynamic and flexible orchestration in the most basic physiological processes, in the plasticity of development, or in the maintenance of form.

If we ask, What lies at the basis of the animal existence? I can't find a better answer than the one expressed in the old adage "It's turtles all the way down." In a living being, there is activity all the way down.

In other words, animals embody creative agency. The understanding of agency, that animals are activities and not entities, that we are dealing with creatively expressive beings and not bio-machines, leads to new and surprising insights into the nature of development and evolution. Theories about development and evolution will remain inadequate as long as they ignore the fundamental role of the dynamic life of beings.

Part One offers portraits of six different mammals. In each I strive to show the unity of the animal, but each also has a somewhat different emphasis.

I begin with an animal that caught my attention because it is so strange—the three-toed sloth. Compared to monkeys that also dwell in the trees of tropical Central and South America, the sloth is a decidedly slow and withdrawn mammal—in movement, digestion, and reactions. Its passive nature is revealed in many remarkable anatomical and physiological characteristics. But the activity it seems to lack also allows the sloth to host a whole array of other creatures in its fur—algae, moths, mites, and more. It becomes a steady, slowly moving, and teeming microhabitat within the rainforest. Almost merging with the vegetative life of its environment, the sloth brings an "ever-smiling," plantlike slowness into its world.

The elephant is an animal we are all familiar with. The more I studied details of the elephant's trunk and all it can do, the more I realized that this organ is not just a remarkable kind of fifth limb emerging from the head; it is also a revelation of qualities you find in many other features of the animal. The flexible, powerful, and multifunctional trunk provided a key to understanding characteristics of the elephant that speak a unified language. Everywhere we look—in anatomy, physiology, or behavior—we discover the features of sensitive flexibility wedded with power and strength.

A short chapter on the star-nosed mole provides a glimpse into the life of a little-known animal. I address the question "How does an animal experience the world?" The star-nosed mole orients itself mainly through a highly refined sense of touch, while various other senses recede. How can we imagine the world from a mole's perspective?

We can certainly say that physically an American bison (popularly known as the "buffalo") ends at the boundaries of its large and distinct body. But as a living animal, in its ecological relations and in its behavior and interactions with other animals it extends far beyond that boundary. This "peripheral bison" participates actively in the larger environment that supports it and that it also molds. Before its near extermination in the nineteenth century, the bison also extended into the life of the Plains Indians, who lived from the bison, both physically and spiritually. This chapter reveals the expansive and many-layered nature of an iconic symbol of North American wildlife.

As prey and predator, the lives of zebra and lion are deeply inter-twined. The zebra shows stamina in all its activities—grazing, grinding, digesting, standing, running. Such stamina and constancy is a central aspect of the language of the zebra. Endurance is the physiological and behavioral expression of the stability revealed in its skeletal anatomy. In contrast, the lion lives in its muscles, which function through an interplay between tension and relaxation. The life of a lion oscillates between extremes—focused, powerful action in the hunt, followed by complete relaxation and lassitude. These starkly different ways of being illuminate each other, just as they are dependent on one another.

The chapter on zebra stripes marks the transition to Part Two. In the chapters on zebra stripes, the giraffe, and the frog, I embed the animal portrayals within the larger topic of evolution and explanatory theories. In my view, explanatory theories often hinder us from seeing the complexity and riddles that animal life presents. Theories become lenses we forget to take off, and therefore we get stuck in viewing animals (and life in general) in narrow and inadequate terms. My aim is to open up broader perspectives rather than to nail things down.

In the zebra and giraffe chapters, I show the shortcomings of trying to explain characteristics (stripes or long neck) in isolation from the animals as complex, integrated wholes. The widespread attempt to explain "traits" in evolutionary biology ignores the unitary nature of animals. As a result, the explanations always fall short. A holistic understanding of organisms is what provides a solid basis for thinking about evolution.

At least 19 different "explanations" have been proffered to answer the question "Why do zebras have stripes?" All unsatisfactory, because they are not only highly speculative, but also based on the unbiological notion that such a complex feature of an animal could be explained in a purely utilitarian way.

The giraffe with its long neck has often been used to illustrate evolutionary theory. One separates in thought the neck from the rest of the animal and tries to explain it. What emerges is a caricature, not an animal. As a counterbalance to this approach, I present a detailed holistic picture of the giraffe that considers the many facets of its unique gestalt, physiology, and behavior. Based on this portrayal, I consider giraffe evolution. Once we take the active and integrated character of organisms seriously, we can no longer view them as passive players in the drama of evolution.

Do frogs come from tadpoles? The question is not as stupid as you might think. In fact, there are justifiable reasons to say that an adult frog does not come from a tadpole. (I won't give them away here.) Based on a careful consideration of frog metamorphosis and frog fossil history, I come to see development and evolution as actual creative activity. Organisms are agents in their own becoming and this insight has radical implications for our understanding of developmental and evolutionary processes.

In Part Three I embed the portrayal of the cow—a domesticated animal—within the question of human responsibility. After painting a picture of the dairy cow and its treatment in modern industrial agriculture, I highlight the challenges we face as human beings in taking responsibility for these animals whose evolution we have so strongly influenced and guided. We are intertwined with them, and they have become dependent on us. How do we work with this responsibility?

In the concluding chapter, I elaborate on what I mean by a "biology of beings," what carrying out the approach entails, and what its implications are for our relation to animals and the natural world more broadly. We can become sensitive to what is creatively at work in each and every creature only insofar as we bring its being alive in us. As Goethe wrote, "If we want to behold nature in a living way, we must follow her example and become as mobile and flexible as nature herself."[8] By entering into the living qualities of nature we will be all the more able to further, rather than disrupt, the health of the planet. And our world depends on that.

PART 1

Portraits

CHAPTER 1

In Praise of Slowness
What Does It Mean to Be a Sloth?

One more defect and they could not have existed.

~ George Louis Leclerc, Comte de Buffon

*We conceive of the individual animal as a small world,
existing for its own sake, by its own means. Every creature
is its own reason to be. All its parts have a direct effect on
one another, a relationship to one another, thereby constantly
renewing the circle of life; thus we are justified in considering
every animal physiologically perfect.*

~ Johann Wolfgang von Goethe

The first sloths I encountered were in zoos. They were definitely
not animals that attracted many visitors. But I had heard and read some
about them, so I was interested in seeing them. I was with my family, and
we stood for a while and watched a ball of fur hanging from a limb. It
didn't move. My children soon became impatient—why were we spending time with this animal?—and so I moved on with them, while my
wife stayed a bit longer. Soon we heard her voice, "It's moving!" We ran
back, only to find it stationary again. It didn't move as long as we stayed.
So after some time, we moved along. This and similar experiences only
heightened my interest in this strange mammal.

I had been learning and teaching about animals as highly integrated
organisms. What is the secret to this decidedly passive animal? Is the
sloth, as eighteenth-century natural philosopher Buffon claimed, a defective creature? I had a difficult time thinking that could be the case. But
is it "physiologically perfect" in Goethe's sense? That was hardly obvious. So a journey into the whole-organism biology and ecology of the
sloth ensued.

In the Rainforest

When you walk in a tropical rainforest, you may startle some monkeys and glimpse them scampering away among all the leaves and branches in the canopy above. You will also certainly hear and see birds. But even if you look hard and make lots of noise, you would most likely not startle or see the most prevalent tree-dwelling mammal in the Central and South American rainforests. The three-toed sloth will likely remain still and well concealed in the tree canopy.

To appreciate the sloth, we need to have a sense of the tropical rainforest. It is an ecosystem characterized by constancy of conditions. The length of day and night during the year varies little. On the equator there are 12 hours of daylight and 12 hours of night for 365 days a year. The sun rises at 6 am and sets at 6 pm. Afternoon rains fall daily throughout much of the year. The air is humid (over 90 percent) and warm. The temperature varies little in the course of the year, averaging 77°F (25°C).

Except in the uppermost part of the canopy, it is dark in the rainforest. Little light appears on the forest floor. The uniformity of illumination, warmth, and moisture—in intensity and rhythm—marks the rainforest. And it is hard to imagine a rainforest dweller that embodies this quality of constancy more than the sloth. From meters below, the sloth is sometimes described as looking like a clump of decomposing leaves or a lichen-covered bough. The sloth's hair is long and shaggy, yet strangely soft. The fur is brown to tan and quite variable in its mottled pattern. Especially during the wettest times of year, the sloth is tinted green from the algae that thrive on its pelage, which soaks up water like a sponge.[1]

Since the sloth moves very slowly and makes few noises, it blends into the crowns of the rainforest trees. It took researchers many years to discover that up to 700 sloths may inhabit one square kilometer of rainforest.[2] Only 70 howler monkeys inhabit the same area.

The sloth spends essentially its whole life in the trees. It sits nestled in the forks of tree branches (mostly) or hangs from branches by means of its long, sturdy claws (sometimes). With its belly directed heavenwards, the sloth's fur has a part on the mid-belly and the hair runs toward the back, unlike terrestrial mammals in which the part is on the mid-back and the hair runs toward the belly.

The sloth moves slowly through the forest canopy—rarely a few hundred feet in 24 hours. On average, sloths have been found to move during five to ten hours of the 24-hour day.[3] The remaining time they are essentially still and in one place. ("Resting" is the term often used to describe the sloth's inactive periods, but this expression is not actually appropriate for the sloth. From what activity is it resting?)

FIGURE 1.1. A three-toed sloth (*Bradypus variegatus*) in Brazil feeding in a *Cecropia* tree.

You are most likely to get a good view of a sloth where the forest borders an open area—along a waterway or next to a road. Instead of being surrounded by vegetation you can peer at the forest edge; you are outside looking in. I was in a small canoe-like boat with a group on a tributary to the Amazon River during the high-water season. It was early morning, near a small village, and we were passing by a number of *Cecropia* trees—a favorite species of sloths—surrounded by much open water. Since the water level was so high, we were at the height of the lower crown. To our fortune we saw a couple of three-toed sloths comfortably situated on limbs.

We observed one of the sloths for quite a while (not an issue from the sloth's perspective—it was more up to us how long we could remain active observers seeing so little happen). The sloth clearly noticed us, turning its eternally smiling face from time to time in our direction. It then began to move upward in the crown. This was stunning (at least for me). It was if we were watching a living fluid flow upward, so gentle and graceful were every arm, leg, neck and head movement. There was nothing jagged, jarring, or abrupt. At one moment it released its arms from the branch and slowly lowered its head and upper body until it hung down vertically.

It stayed in this position for a short time. Just as slowly, and seemingly without effort, it raised its head and upper body again and clawed onto the branch. It remained hanging from three claws until we left it.

Clinging in Tree Crowns

The sloth's ability to cling to branches and to hang from them for extended periods of time is related to its whole anatomy and physiology. About the size of a large domestic cat (weighing around eight to ten pounds), the sloth has very long limbs, especially the forelimbs. When hanging, the sloth's body appears to be almost an appendage to the limbs. Feet and toes are hidden in the fur. Only the long, curved and pointed claws emerge from the fur. The toe bones are not separately movable, being bound together by ligaments, so that the claws form one functional whole, best described as a hook (Figure 1.2).[4]

FIGURE 1.2. Two different three-toed sloths in typical "resting" positions.

The two different genera of sloths are named according to the number of claws they possess: the three-toed sloth (*Bradypus*) has three claws on each limb; the two-toed sloth (*Choloepus*) has two claws on the forelimb and three on the hind limb. (There are many differences in detail between these two genera of sloths. I am focusing here primarily on the three-toed sloth.[5])

With its long limbs the sloth can embrace a thick branch or trunk, while the claws dig into the bark. But the sloth can also hang by just its

FIGURE 1.3. A three-toed sloth flowing up a tree limb and then hanging from its short hind legs.

claws on smaller branches, its body suspended in the air. A sloth can cling so tenaciously to a branch that researchers resort to sawing off the branch to bring the creature down from the trees.

All body movements, or the holding of a given posture, are made possible by muscles rooted in the bones. Muscles work by means of contraction. While clinging, for example, some muscles in the limbs—the retractor muscles—are contracted (think of your biceps) while other muscles— the extensor muscles—are relaxed (think of your triceps). When a limb is extended (when the sloth reaches out to a branch) the extensor muscles contract, while the retractor muscles relax. All movement involves a rhythmical interplay between retractor and extensor muscles.

It is revealing that most of a sloth's skeletal musculature is made up of retractor muscles.[6] These are the muscles of the extremities that allow an animal to hold and cling to things and also to pull things toward it. The extensor muscles are smaller and fewer in number. In fact, significant extensor muscles in other mammals are modified in the sloth and serve as retractor muscles. A sloth can thus hold its hanging body for long periods of time. It can even clasp a vertical trunk with only the hind limbs and lean over backward 90 degrees with freed forelimbs. (Just imagine doing this yourself and your belly muscles will already start to ache!)

At home as it is in the trees, the sloth is virtually helpless on the ground. Lacking necessary extensor muscles and stability in its joints, a sloth on the ground can hardly support its weight with its limbs. Researchers know little about natural terrestrial movement of sloths. Normally, a sloth will move from tree to tree up in the crowns, but in areas where roads divide parts of the forest, they may have no alternative but to cross the road. My brother was in Costa Rica and stopped his car to let a sloth, very slowly, drag itself across the road (see Figure 1.4). He stayed there for about five minutes to make sure other cars wouldn't run it over. A sad sight, but the sloth made it. With its limbs splayed to the side, it found holds with the claws of its forelimbs and gradually pulled itself forward, using its strong retractor muscles.

Since the sloth's main limb movements involve pulling, and the limbs do not carry the body weight, it is truly a four-armed and not a four-legged mammal. The hands and feet are essentially a continuation of the long limb bones, ending in the elongated claws, and do not develop

FIGURE. 1.4. A three-toed sloth in Costa Rica dragging itself across a road. (Photo by Tom Holdrege, used with permission)

as independent, agile organs as they do, say, in monkeys. We can also understand the dominance of the retractor muscles from this point of view. The human being, in contrast to most mammals, has arms as well as weight-bearing legs. The arms are dominated by retractor muscles, while the legs have more extensor muscles. Moreover, the arm muscles that move the arm toward the body are stronger than the antagonistic arm muscles that move the arms away from the body. This comparison shows us that the tendency inherent in the arm—the limb that does not carry a body's weight—permeates the anatomy of the sloth.

A sloth becomes quite agile if the forces of gravity are reduced, as in water. In water a body loses as much weight as the weight of the volume of water it displaces (Archimedes' Law). The body becomes buoyant, and in the case of the sloth, virtually weightless. As science writer Fiona Sunquist notes,

> Remarkably, sloths are facile swimmers.... They manage to move across water with little apparent effort. Where the forest canopy is interrupted by a river or lake, sloths often paddle to new feeding grounds. With no heavy mass to weigh them down, they float on their buoyant, oversized stomachs.[7]

With its long forelimbs the sloth pulls its way through the water, not speedily, but in a "beautifully easy going" manner.[8]

On the whole, sloths have little muscle tissue. Muscles make up 40 to 45 percent of the total body weight of most mammals, but only 25 to 30 percent in sloths.[9] One can understand how the reduction of weight in water allows them to be less encumbered in movement. Sloth muscles also react sluggishly, the fastest muscles contracting four to six times slower than comparable ones in a cat. In contrast, however, a sloth can keep its muscles contracted six times longer than a rabbit.[10] Such anatomical and physiological details reflect the sloth's whole way of being—steadfastly clinging in a given position, only gradually changing its state.

The tendency to the reduction of muscle tissue can also be found in the head. There is a reduction in the number and complexity of facial muscles.[11] Through the facial markings the sloth has an expressive face—it appears content and smiling. But this expression is a fixed image, rather than an active expression through movement, since the facial area itself is relatively

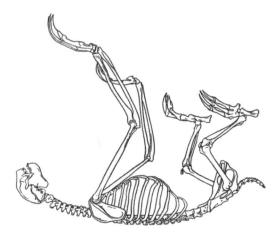

Figure 1.5. Skeleton of a three-toed sloth. (Adapted from de Blainville,1840; also reproduced in Young, 1973, p. 600)

immobile. The outer ears are tiny and are essentially stationary. The sloth alters the direction of its gaze by moving its head, not its eyeballs. This rather fixed countenance is dissolved at the lips and nostrils, which, as the primary gateways to perceiving and taking in food, are quite mobile.

Laxity

If we look for the embodiment of fixed form in the organ systems of a mammal, we come to the skeleton. The bony skeleton gives the mammal its basic form and is the solid anchor for all movement. The limb bones develop their final form in relation to both gravity and their own usage. An injured quadruped mammal will lose bone substance in the leg it is not using, which does not carry any weight. Conversely, in the other three limbs bone matter is laid down to compensate for the increase in weight carried and muscular stress.

Hoofed mammals like zebras or giraffes spend most of their lives standing and walking. By virtue of their skeletal architecture they can relax their muscles and even sleep while standing. Their legs are solid, stable columns that carry the body's weight. Active arboreal mammals, like monkeys, have, of course, nothing of the skeletal rigidity of ground-dwelling quadrupeds. They have flexible joints and muscular agility that allow for actively

Figure 1.6. Three-toed sloth. Note the orientation of the head.

swinging, jumping, and grasping. Their long tail serves as an extra limb for balance. They have an energetic lightness in movement and expression.

Then there is the sloth. Spending most of the day reclining on branches or hanging, the sloth gives itself over to gravity rather than resisting it and living actively within it. A sloth kept on the ground in a box developed raw feet from the unaccustomed pressure.[12]

The sloth has very loose limb joints. In his detailed study of the limbs of the two-toed sloth, Frank Mendel points out the unusual nature of the "poorly reinforced and extremely lax joint capsules."[13] This anatomical peculiarity allows a wide range of limb movement and is connected with the fact that the joints are not subject to compression as they are in weight-bearing limbs. Through clinging and hanging, the joints of a sloth are being continually stretched. Similarly, the sloth has a flexible, curved spine. The hoofed mammal, in contrast, has a stiff, straight spine, from which the rib cage and internal organs of the torso are suspended. A zebra would be as ungainly in a tree as a sloth is on the ground.

The laxity of the joints finds positive expression in the flexibility and fluidity of slow movement that the sloth embodies. A sloth can twist its forelimbs in all directions and roll itself into a ball by flexing its vertebral column. It can move its head in all directions, having an extremely flexible neck. Imagine a sloth hanging from all four legs on a horizontal branch. In this position the head looks upward (like when we lie in a hammock). Now the sloth can turn its head—without moving the body—180 degrees to the side and have its face oriented downwards. As if this were not enough, the sloth can then move its head vertically and face forward—an upright head on an upside down body (Figure 1.6)! When it sleeps, a sloth can rest its head on its chest.

The sloth's neck is not only unique in its flexibility, but also in its anatomy. Mammals have seven neck (cervical) vertebrae. Even the long-necked giraffe and the seemingly neckless dolphin—to mention the extremes—both have seven cervical vertebrae. This mammalian pattern is abandoned by only the sloth and the manatee. The three-toed sloth often has eight or nine neck vertebrae, some of which occasionally fuse into one. The two-toed sloth usually has six cervical vertebrae. In both genera these neck "oddities" are accompanied by other changes in the spine, such as lumbar vertebrae that bear rudimentary ribs.[14]

It's interesting in this connection that three-toed sloths also show remarkable variation between individuals in the semicircular canals of the inner ear.[15] These canals are connected with the sense of balance as it is related to the head and its movements in different directions. In other species, such as squirrels or moles, there is much less variation between individuals. In the sloth the shape of the three canals, their thickness, and the angles between them—which in other animals hardly vary—vary widely. So here also, as in the neck, the sloth is less constrained or rigid in the shaping of an otherwise highly defined organ. The vestibule of the inner ear, which is also connected with the sense of balance, and the cochlea, which is connected with hearing, also show considerable variation, but not to the degree of that in the semicircular canals. What all this variability signifies for the sloth in its activities is unknown.

It is as if the sloth were saying, anatomically, "Does it really matter so much?" For the life it lives, evidently not.

Drawing In

The head is the center of the primary sense organs through which an animal relates to its environment. A sloth's eyesight is poor, and it is shortsighted.[16]

The eyes lack the tiny muscles that change the form of the lens to accommodate for changing distances of objects. As if to emphasize the unimportance of its eyes, the sloth can retract them into the eye sockets. The pupils are usually tiny, even at night. Clearly, a sloth does not actively penetrate its broader environment with its vision, as do most arboreal mammals like monkeys.

The outer ears are tiny and hardly visible on the head. While a female sloth clearly responds to the cries of her offspring, cries of a hawk flying overhead or the sound of a gun being fired can be wholly ignored (or not noticed?) by a sloth.[17] It is the drawing-in senses of smell and taste that dominate in the sloth (see the next section).

The head is relatively small and appears almost as a broadened neck. The first cervical vertebra (the so-called atlas) is nearly as wide as the widest part of the skull. The skull itself is rounded and self-contained— superficially resembling a monkey's skull more than a grazing herbivore's (Figure 1.7). Most herbivores have an elongated snout that they use as a limb—standing as they do on all four legs—to reach their food. The sloth's forelimbs have this function and thus its snout is short. The premaxillary bones—important in forming the elongated mammalian

Three-toed sloth

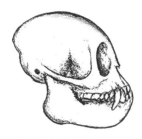

Monkey (*Presbytis*; southeast Asia)

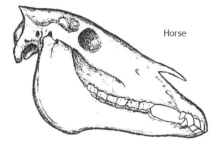

Horse

FIGURE 1.7. Skulls of a three-toed sloth (top left), a leaf monkey (*Presbytis*, top right), and a horse. The skulls are not drawn to scale.

snout—are tiny in the sloth. Moreover, the upper jawbones (maxillae) and the nasal bones are also short in the sloth. The sloth's skull does not project forward. Here we see the wonderful congruence of anatomy and behavior—the unity of an animal that recedes from active engagement.

The expression of pain is a barometer for the way an animal experiences its own body in relation to the environment. Pain is one way an animal experiences the external world penetrating and harming its biological integrity. Here's an example from a family that kept a sloth at their home in Brazil:

> 'Sloth burning!' … we leap to our feet and run frantically round trying to discover where [the sloth] has fallen asleep. On the kitchen stove? No! On the water heater in the bathroom? No! There he is on top of the floor lamp in the drawing-room, with his bottom touching the big electric bulb!... We struggle to get him down, but he clings desperately to his perch, refusing to budge and protesting with many ah-eees against our unwarranted disturbance of his slumbers.[18]

A burning behind and no response! Sloths are reported to "survive injuries that would be deadly within a short time to other mammals."[19] "I have known a sloth to act normally for a long time after it had received a wound which practically destroyed the heart."[20] These examples show that the sloth does not seem to notice such intrusions of its boundaries and continues to live despite them. It is somehow withdrawn; its form of embodiment is not one of sensitive reactive presence (think monkeys).

Feeding in a Sea of Leaves

Moving unhurriedly through the crown of a tree, the sloth feeds on foliage. We usually think of leaf browsing in connection with ground-dwelling mammals, such as deer. There are, in fact, only ten species of mammals that are specialized arboreal leaf eaters.[21] Leaves are an abundant and constant source of food, and plants need not be chased down. Spending virtually its entire life in the crown of trees, a sloth is literally embedded in and surrounded by its food at all times and in all directions. Tropical trees do lose their leaves, but usually not all at once. Sometimes one and the same tree may lose its leaves on one branch, while it sprouts new ones on others. Sloths don't eat just any leaves. They seem to prefer younger leaves, and each individual animal has its own particular repertoire of about 40 tree species from which it feeds.[22] A young sloth feeds together with its mother, often licking leaf fragments from the mother's lips. After its mother departs, leaving it at the age of about six months on its own, the young sloth continues to feed from those species it learned from its mother.

This specificity is probably a major factor in the inability to keep three-toed sloths alive in zoos. They usually die of starvation after a short period of time. In contrast, the two-toed sloth is more flexible and survives well in captivity, eating assorted fruits and leaves.

In finding and assessing food, the sloth relies not so much on vision as on a sense that entails drawing the environment into itself: the sense of smell. Naturalist William Beebe gives a nice description:

> I placed a sloth, hungry and not too disturbed, on an open area under the bamboos, and planted four shoots twenty feet away in the four directions of the compass. One of these was *Cecropia* [a primary food of three-toed sloths] camouflaged with thin cheesecloth, so that the best of eyesight would never identify it, and placed to the south, so that any direct wind from the east would not bring the odor too easily. The sloth lifted itself and looked blinkingly around. The bamboos thirty feet above, silhouetted against the sky, caught its eye, and it pitifully stretched up an arm, as a child will reach for the moon. It then sniffed with outstretched head and neck, and painfully began its hooking progress toward the *Cecropia....* Not only is each food leaf tested with the nostrils, but each branch.[23]

So we need to imagine the sloth orienting itself in a sea of wafting scents.

When a sloth is in the immediate proximity of leaves it feeds on, it slowly extends its forelimb or hind limb, hooks the branch with the claws, and brings the leaves to its mouth. Remarkably, sloths can feed in all positions, even hanging upside down. A young captive two-toed sloth showed "decided preference for eating upside down in the manner of adult sloths."[24]

Having no front teeth (incisors or canines), it tears off the leaves with its tough lips. While feeding, the sloth continuously chews with its rear, peg-like teeth, and simultaneously moves food backward with its large tongue in order to swallow.

Unlike most leaf-eating mammals, the sloth lacks many deeply rooted, hard, enamel-covered, and distinctly formed grinding teeth—enamel is the hardest substance a mammal can produce. The sloth also has comparatively few teeth (18 compared to 32 in most deer). Lacking enamel, the teeth wear easily, but in compensation they grow slowly throughout the animal's life. Growth and wear are in balance. The sloth's teeth emerge as simple cones and take on a characteristic form in the course of life. [25]

That the teeth lack enamel and allow wear and tear to form them is not at all characteristic for a mammal. Usually mammal teeth emerge highly formed, and scientists can determine a species by the tooth characteristics. But "since sloth teeth acquire their individual characteristics through wear, it is very difficult to distinguish the young of one genus from those of another based upon shape or location of dentition."[26]

The softness and lack of distinct, species-determined tooth form indicate a kind of anatomical passivity, if I may call it that. It seems to be a signature characteristic of the sloth to let itself be so strongly formed through the environment—whether in tooth form or the green tint of its fur. We will discover more features of this nature.

Centered in its Stomach

Digestion in the sloth occurs at a remarkably slow rate. In captive animals "after three or six days of fasting the stomach is found to be only slightly less full."[27] Leaves are hard to digest and not very nutrient-rich, consisting primarily of cellulose and water. Only with the help of microorganisms in the stomach can the sloth digest cellulose, breaking it down into substances (fatty acids) that can be taken up by the blood stream.

The sloth's stomach is—remarkably—multichambered, somewhat resembling that of ruminants (such as bison or cows), and is clearly the center of the digestive process. The stomach is enormous relative to the animal's overall size. It takes up most of the space of the abdominal cavity and, including contents, makes up from 17 to 37 percent of total body weight.[28] Food stays a long time in the stomach, and it takes a long time to digest the leaves. On the basis of field experiments, researchers Montgomery and Sunquist estimate that it takes food about ten times longer to pass through a sloth than through a cow. Moreover, the sloth also digests less of the plant material than most other herbivores. [29]

Through its stomach a mammal senses hunger. Most grazing mammals spend a large part of their time eating, so that food is continuously passing through their digestive tract. The sloth is, once again, an atypical herbivore since it feeds for a comparatively small portion of its day—one to four hours of a 24-hour day.[30] A small rainforest deer, the same size as a sloth, ate six times as much during the same period of time.[31] The howler monkey, which also lives in the canopies of Central and South American rainforests and whose diet comprises only about 50 percent leaves, eats about seven times as many leaves as do sloths. With its slow metabolism and digestion, the sloth's stomach remains full, although the animal eats so little.

As a stark contrast, we can think of carnivores like wolves or lions that regularly, as a normal part of their lives, experience empty stomachs. Their hunting drives are directly connected to their hunger. Hunger brings about the maximal aggressive activity of these animals. When a lion has gorged itself on 40 pounds of meat, it becomes lethargic and sleeps for an extended period of time. The sloth's constantly full stomach is a fitting image for its consistently slow-paced, phlegmatic, and nonaggressive way of life.

After about a week of feeding, sleeping, and external inactivity, a change occurs in the sloth's life. It begins to descend from its tree. Having reached the forest floor, it makes a hole in the leaf litter with its stubby little tail. It then urinates and defecates, covers the hole, and ascends back into the canopy, leaving its natural fertilizer behind.[32]

The feces decompose very slowly. The hard pellets can be found only slightly decomposed six months after defecation. Normally, organic material decomposes rapidly in the warm and moist conditions of the rainforest. For example, leaves decompose within one to two months (a process that can take a few years in a temperate-climate forest). Ecologically, sloth feces "stands out as a long-term, stable source [of nutrients] ... and may be related to stabilizing some components of the forest system.... Sloths slow the normally high recycling rates for certain trees."[33] Sloths contribute not only slow movement to the rainforest but slow decomposition as well! How fitting.

It is estimated that a sloth can lose up to two pounds while defecating and urinating, more than one-fourth of its total body weight.[34] If one imagines a sloth with a full stomach (which it always seems to have) just prior to excreting, then more than half of its body weight is made up of its food, waste, and digestive organs! This quantitative consideration points to the qualitative center of gravity in the animal's life. But the sloth's stomach is more like a vessel that needs to remain full than a place of intensive muscular activity, secretion, mixing, and breaking down, as it is, for example, in the cow.

Stretching Time

"Sloths have no right to be living on this earth, but they would be fitting inhabitants of Mars, whose year is over six hundred days long."[35] In his typical humorous way, naturalist William Beebe expressed with these words his awe of the sloth's ability to "stretch" time, another way of characterizing their slowness. We have seen how this quality permeates every fiber of their day-to-day existence. It is therefore not so surprising to find that the development of sloths takes a long time.

Three-toed sloths have a gestation period of around six months, compared to a little over two months in the similar-sized cat, while two-toed sloths give birth only after eight to ten months.[36] Initially more surprising was the rediscovery of a female sloth in the rainforest 15 years after she had been tagged when she was already an adult. This means she was at least 17 years old, "an unusually long life span for such a small mammal."[37] Thus, regarding time, the qualities of the sloth certainly speak a unified language.

Going with the Flow

Since sloths are externally inactive a good portion of the 24-hour day and the remaining time is spent slowly moving and feeding, they perform about 10 percent of the physiological work of a mammal of similar size.[38] All metabolic processes are markedly measured in tempo and intensity. The basal metabolic rate, which is a measure of how much energy an animal uses to carry out its life functions in a resting state, is about 40 percent less in three-toed sloths than in other mammals of comparable size.[39] Sloths use little oxygen, breathe slowly, and the respiratory surface of their lungs is small. While an inactive sloth takes on average nine breaths per minute, a cat takes about 20, and breathing out takes up about 70 percent of the time of one in-breath and out-breath cycle.[40]

All metabolic activity produces warmth. Warmth is also needed for activity, for example, in the exertion of muscles, which in turn results in more warmth production. Birds and virtually all mammals not only produce warmth, but also maintain a constant body temperature. This is a striking physiological feat. A warm-blooded (endothermic) animal is like a radiating, self-regulating center of warmth. Warmth constantly permeates the whole organism.

Most mammals maintain a constant core body temperature of about 98°F (36°C), which changes only little despite variations in environmental temperatures. For example, in a laboratory experiment a mouse's internal temperature changes only four-tenths of one degree Celsius when the outer temperature rises or falls twelve degrees.[41] A sloth does not create so much warmth, and its body temperature can vary markedly.

During the morning, as the ambient temperature rises, its body temperature rises. When found on sunny days, sloths are often on an outer branch, belly-side up and limbs extended, basking in the sun. Body temperature usually peaks at about 96–100°F (36–38°C) soon after midday.[42] It then begins to fall, reaching a low point of about 86–90°F (30–32°C) in the early morning. The body temperature is usually about 12–18°F (7–10°C) higher than the ambient temperature. Interestingly, sloths eat more when the ambient temperature is higher, unlike many mammals that eat more when it is cold.[43]

Although sloths are often active at night, their body temperature does not rise with their increased activity. This shows, in contrast to other mammals, that the sloth's body temperature is less affected by its own activity than by the ambient temperature. According to scientist Brian McNab, the sloth "almost appears to regulate its rate of metabolism by varying body temperature, whereas most endotherms [warm-blooded animals—mammals and birds] regulate body temperature by varying the rate of metabolism."[44]

Beneath its outer hair, a three-toed sloth has an insulating coat of fur comparable to that of an arctic mammal, which seems at first rather absurd for a tropical animal. It has, like an arctic fox, an outer coat of longer, thick hair and an inner coat of short, fine, downy fur. These allow the sloth to retain the little warmth it creates through its metabolic processes. But, characteristically, the sloth cannot actively raise its body temperature by shivering as other mammals do. Shivering involves rapid muscle contractions that produce warmth.

Clearly, the sloth is at home in the womb of the rainforest. Not only by virtue of its coloring and inconspicuous movements does the sloth blend into its environment, but also through its slowly changing body temperature.

Sloth as Habitat

As if to emphasize its passive, somewhat withdrawn character, the sloth becomes a habitat for myriad organisms. I have mentioned the algae that live in its fur, giving the pelage a greenish tinge. In addition to the usual ticks and flies that infest the skin and fur of other mammals, a number of sloth-specific moth, beetle, and mite species live on the sloth and are dependent upon it for their development.[45] The sloth moths and beetles live as adults in the sloth's fur. Some species live on the surface, and others inhabit the deeper regions of the fur. They are evidently not parasitic; their source of food is unknown.

When the sloth descends from a tree to defecate and urinate, female moths and beetles fly off the animal and lay their eggs in the sloth's dung. The wings of one moth species break off soon after they inhabit the sloth, and they crawl around deep in the sloth's fur. Evidently, the sloth's relatively long period of defecation, which lasts a few minutes, gives the insects the time they need to make their journey off the sloth. In this way the slowness of the sloth serves these most "slothful" of sloth moths!

Larvae develop out of the eggs and then feed on the dung (which, you remember, decomposes slowly). The larvae pupate in the dung and the winged adult moths (or beetles) fly off to inhabit another sloth. Various species of insects and mites inhabit any given sloth, and the numbers of specimens of each species varies greatly, ranging from a few to over a hundred. In one single sloth 980 beetles of a particular species were found.

The sloth has been observed grooming its fur. This is typical mammalian behavior and does rid an animal of some of its "pests." From this utilitarian point of view, the sloth's grooming is not very effective. Typically, sloths groom slowly, and sloth moths "may be seen to advance in a wave in front of the moving claws of the forefoot, disturbed, but by no means dislodged from the host."[46] Its measured pace of life, its unique excretory habits, and the consistency of its dung allow the sloth to be a unique habitat for such a variety of organisms.

One research team has investigated the ecosystem of the sloth fur in more detail and discovered that sloths lick their fur and ingest some of the algae living there.[47] The algae are more easily digestible and have a higher lipid content than the leaves sloths primarily feed on. So sloths are "grazing the 'algae gardens'" that their fur lets grow. The "algae gardens," in turn, are evidently fertilized by nutrients provided by their insect inhabitants (through urine, feces, and decomposing body parts). So sloth fur is a kind of a microcosm of the larger ecosystem of which it is a part.

Other animals haven't overlooked that sloths are a habitat teeming with life. For example, researchers observed two birds—yellow-headed caracaras—"picking from the fur of the sloth.... The sloth showed no sign of defensiveness or aggression toward the caracaras even when they were foraging on its head and neck. It assumed a relaxed posture, reclining on a branch with its front legs extended behind its head. The behavior continued for 5-10 min[utes]."[48] Why not let birds do a bit of feeding and grooming?

Sloth, tree, moth, beetle, bird, and algae all interweave. On the one hand, each of these organisms—as a center of form and activity—is itself, is its own being. On the other hand, the existence of each is bound up with and made possible by myriad other creatures and, of course, light, warmth, air, and rain. Because every organism is itself by virtue of others, there are no hard and fast boundaries in the living world. In the chapter on the bison I will look more closely at the question of where an animal ends.

Encircling the Unspeakable: The Animal as a Whole

I'd like to return to the statements quoted at the beginning of this chapter. George Louis Leclerc, Comte de Buffon, was a well-known eighteenth-century French natural philosopher (as scientists were called in those days) who studied many animals, among them the sloth. He came to the conclusion that "one more defect and they could not have existed."[49] The sloth's remarkable characteristics were for him defects. And they are, if you take the point of view of a horse, eagle, jaguar, or human being.

But William Beebe had a great retort to Buffon: "A sloth in Paris would doubtless fulfill the prophecy of the French scientist, but on the other hand, Buffon clinging upside down to a branch of a tree in the jungle would expire even sooner."[50] Beebe recognized how each creatures has its unique manner of existence. Writing some years after Buffon's death, the poet and scientist J. W. von Goethe articulated such an integrative view of animals:

> We conceive of the individual animals as a small world, existing for
> its own sake, by its own means. Every creature is its own reason to be.
> All its parts have a direct effect on one another, a relationship to one
> another, thereby constantly renewing the circle of life; thus we are
> justified in considering every animal physiologically perfect.[51]

When Goethe calls an animal "perfect," he means that each animal has its own unique way of being—its specific integrity that we can try to understand. But this is no simple matter. Goethe recognized that "to express the being of a thing is a fruitless undertaking. We perceive effects and a complete natural history of these effects at best encircles the being of a thing. We labor in vain to describe a person's character, but when we draw together actions and deeds, a picture of character will emerge."[52]

In this portrayal of the sloth, I have discussed many details, because through them the whole lights up. Henri Bortoft puts it well when he says, "The way to the whole is into and through the parts. The whole is nowhere to be encountered except in the midst of the parts."[53] The emergent picture of the whole does not and cannot encompass the totality of its characteristics and interactions. We can always discover new details. It is not about knowing all the facts, but about seeing them in relation to one another and as expressions of the wholeness of the animal. Such understanding hinges on our ability to overcome the isolation of separate facts and to begin to fathom the animal as a whole, integrated organism—which includes its relations to other creatures, substances, and forces that support its existence.

The whole is elusive and yet, at every moment, potentially standing before the mind's eye. When we begin to see how all facets of the animal are related to each other, then it comes alive for us. Or, putting it a bit differently, the animal begins to express something of its life in us. Every detail can begin to speak "sloth," not as a name, but as a qualitative concept to which no definition can do justice.

I have tried to describe the sloth in a way that allows us to catch glimpses of its wholeness. I can now refer to such characteristics as slowness, inertia, blending in with the environment, receding or pulling in, and not actively projecting outward. Each expression is a different way of pointing to the same coherent whole. Taken alone, as abstract concepts or definitions, these pointers are empty. They are real only inasmuch as they light up within the description or perception of the animal's characteristics. But they are not things like a bone or an eye. They are, in context, vibrant concepts that reveal the animal's unique way of being.

Let's return to the sloth, up in the crown of a rainforest tree, nestled on a branch. In its outer aspect, it blends in with its environment. The sloth's body temperature rises and sinks with the ambient temperature. There are no sudden or loud movements. Its green-tinged, mottled brown coat lets it optically recede into the wood and foliage of its surroundings. And like the tree bark, the sloth's fur is teeming with life.

The round form of its head is the anatomical image of the sloth's withdrawn relation to its environment. There are no large, movable, reactive outer ears. And the eyes are rarely, if ever, moved. The sloth has no protruding snout. It draws the scents of the environment, especially of the leaves it feeds upon, into its nose. But much of the day the sloth is inactive. Even when awake, the sloth seems not to live as intensely in its body as other mammals, being quite insensitive to pain.

The sloth does not carry its own weight; rather, it clings to or rests on an outer support. Its skeletal system is not characterized by stability, but by looseness. This laxity allows the sloth to adopt positions that would be contortions in other animals. The sloth makes mostly steady pulling movements with its long limbs, a capacity based on the dominance of retractor muscles. With its loose-jointed body, it "flows" when it climbs.

Developing slowly in the womb, the sloth also has a long, slow life. It moves unhurried through the crowns, feeding on the leaves that surround it from all sides, bathing in its food source. The leaves pass through the animal at an almost imperceptibly slow rate. The sloth's stomach is always filled with partially digested leaves—leaves that previously surrounded it in the tree canopy. Even its dung disappears slowly, moderating the otherwise rapid decomposition processes that characterize the warm and humid rainforest climate.

Almost merging with the vegetative life of its environment, the sloth brings—as an animal and ever-"smiling"—a plantlike slowness into its world.

CHAPTER 2

The Flexible Giant
Approaching the Elephant

W<small>HEN</small> I <small>HAVE ASKED</small> students or participants in workshops what makes the elephant stand out—what makes it unique—a list comes together quickly:

- huge body
- long, flexible trunk
- pillar-like legs
- big ears
- long tusks
- grey, almost hairless, but wrinkly skin
- roundish, softly treading feet
- loud trumpeting
- comparatively small eyes
- high intelligence
- complex social behavior

Surely, we can gain greater knowledge of the elephant by studying each of these characteristics in more detail. But in the process we run the constant danger of losing sight of the elephant in all its features. We may find ourselves on a path of endless analysis that leads us further and further away from the elephant itself. Think of the story of the blind men who report on their disparate experiences of the different parts of the animal.

An encyclopedic compendium of facts about the elephant is not what I am looking for. The elephant is not its anatomy, beside its physiology, beside its ecology, beside its behavior; and it is not the sum of them all. The whole is not gained by piecing together parts. It is, rather, the unity of the organism that expresses itself in each one of these facets of its being. Can we learn to see that unity through the careful study of the animal's manifold features? To keep analysis from taking on a life of its own, while gaining knowledge of detail, I continually return to the question, Who are you, elephant? The idea of the coherent animal, framed as a question, becomes the guiding light of inquiry. The challenge is to articulate this unity—to make it visible to our understanding.

The Trunk and Flexibility

It is hard to imagine an organ more flexible than the elephant's trunk. The trunk sweeps to and fro along the ground and then stretches high into the air. It curves up and back, making undulating movements while spraying sand or water over the body. Bending around full circle, it brings food or water to the inconspicuous mouth at its base. When two elephants meet, they often intertwine their trunks with beautiful spiraling movements.

No bone or cartilage hinders this fluid motion. The trunk is muscle through and through. Anatomically, the trunk is the elephant's extremely elongated upper lip and nose. As a nose it contains two nostrils with air passages that extend up into the head and are separated by a wall (septum). While in other mammals this septum consists of cartilage, giving the nose rigidity, the septum in the elephant is muscle, becoming cartilage only at the base where it is rooted in the skull. The trunk can bend and stretch in all directions. The body of the trunk consists of a complex fabric of lengthwise, crosswise, radial, and diagonally spiraling layers of muscle. These layers are in turn differentiated into countless subunits (fascicles) that allow such fine and smooth muscular coordination.[1]

The tip of the trunk is especially dexterous, the elephant using it as we do our fingers for the sensitive exploration of objects, and also to pick up and manipulate them. (The African elephant has a two-tipped trunk, while the Asian elephant's trunk has one tip.) Elephant researcher Joyce Poole observes,

> In studying the behavior of elephants, I have found watching the tip of the trunk to be highly informative. The tip of an elephant's trunk is almost never stationary, moving in whatever direction the elephant finds interesting. An elephant's attention usually is stimulated by what other elephants are doing, and by observing the trunk tip I often have been alerted to subtle behavior that is taking place in the group that I might otherwise have missed.[2]

The physical flexibility of the trunk finds its functional expression in an astounding repertoire: picking, grabbing, enwrapping, reaching, lifting, and pulling—all while gathering food and putting it in the mouth; sucking in and spraying water into the mouth to drink; smelling with probing, searching motions; breathing, including use as a snorkel in water; spraying mud, or sand onto the skin (or onto other elephants in play); caressing, slapping, nudging, lifting, shoving, or trumpeting in social interaction.

Since it can carry out so many activities with one organ, the elephant can at any given moment shift rapidly from one activity to an entirely different one. It drinks, then sprays, then trumpets, then rubs and sniffs, and then pulls down a branch and feeds. This remarkable functional diversity is an expression of the elephant's behavioral plasticity—it doesn't get stuck in its ways. The trunk embodies physical, functional, and behavioral flexibility.

Therefore, it is not surprising that young elephants must learn to use their trunks. Cynthia Moss, who has studied elephants with great care and persistence for decades, describes:

FIGURE 2.1. A group of young male African elephants drinking in the Zambezi River (border of Zimbabwe and Zambia).

> A calf will frequently try to grasp with and manoeuvre its trunk, even a very young one like Ely whose trunk resembled a wobbly, out of control rubber hose. He spent a lot of time wiggling the trunk up and down and around in circles, or sticking it in his mouth and sucking on it. Now the two "fingers" on the tip of the trunk pulled and pushed the stick, until

he finally managed to pick it up. He waved it aloft like a baton and, having accomplished that feat, dropped it and wandered away.[3]

When observing a larger group of elephants it is easy to pick out a young calf, not only because of its size. Its trunk seems to flail about and the end appears limp—it lacks the muscular control and dexterity so evident in the trunk of older animals. The trunk is clearly a tool to be mastered. But just as it never becomes physically rigid, so is the elephant always able to learn new tasks with its trunk.

Variety and Versatility in Food and Habitat

In addition to their staple foods—grasses, bamboos, legumes and the bark of selected plants—elephants have an amazing variety in their menu. Climbers, creepers, palms and succulents are eaten. The leaves of various fig trees are much sought after. Fruits of the tamarind, the wood apple [and] the wild mango are seasonal delicacies…. No other land animal is able to exploit such a wide range of plant resources. This is made possible by that unique organ, the trunk, whose finger at the tip can delicately pick up a tiny object and whose reach extends to a fig tree five meters high.[4]

Such is the menu of a wild-living Asian elephant. And while the trunk is certainly essential to the elephant's ability to access such a variety of food, the versatility it exhibits is in fact inherent in the whole animal. The elephant's great height—increased by the trunk—gives it the ability to reach into trees. It can expand this already large feeding zone even more by shaking a tree, which makes the fruits drop off, or by pushing over a whole tree to get at fruits, leaves, and bark. Here its bulk and strength come into play. The elephant breaks branches with its feet and tears bark off of trees with its tusks.

The mobile lips and tongue are also essential for the deft way it is able to manipulate its food:

The sheaf of grass is placed crosswise in the mouth, with the basal root part projecting from the lips on one side and the tips of the blades on the other, then the projecting parts of the sheaf are bitten through and allowed to fall to the ground, and the rest is

Figure 2.2. An African elephant standing on its hind legs to reach fruits in a tree.

masticated and swallowed. When the grass is tender, the blades are consumed and … the rejected parts of sheaves consist largely of the basal stalks and roots; when the blades are mature and hard, but basal culms are succulent, the apical part of the sheaf is rejected and the culms (with only the roots bitten off) consumed. The placing of the sheaf in the mouth, and the consumption of a part of it, is a selective action and not purely mechanical.[5]

This discrimination in feeding is found in a broader sense as well. Elephants are more selective in the kinds of leaves they will eat than, say, giraffes or antelopes are.[6] At the same time they feed on a much wider variety of plant parts (roots, bark, leaves, stems and fruits).

The food is ground between massive molars, the jaws moving to and fro in rhythmical action. That the elephant can not only access a large variety of food from ground to tree canopy, but also digest this variety, shows that it has, compared to most herbivores, an unspecialized digestive system. With its stomach and intestines the elephant can digest hard grasses, soft leaves of broadleaf trees and shrubs, mineral- and fiber-rich bark, fruits, or flowers, depending on their availability. Given the possibility, the elephant chooses variety.

In his well-known experiments with a captive Asian elephant, Francis Benedict found that the elephant digested only about 44 percent of its food, in contrast to 70 percent for the cow.[7] More recent observations on two African elephants suggest that the elephant may only digest as little as 22 percent of the food it takes in.[8] Food passes through the elephant quickly, staying in the body only for 20 to 46 hours in contrast to six days in the cow. What comes out as dung can still have many intact plant fibers. These observations suggest that although the elephant takes in large quantities of food, it does not put as much physiological activity into digesting the food as does a ruminant like the cow with its four-chambered stomach.

The elephant's ability and tendency to feed on a variety of food is mirrored in its capacity to thrive in a variety of habitats: tropical rainforests, montane forests, all types of savannahs, which are a varied mosaic of trees and grassland, and even the desert. This is another sign of its versatility.

Rainforest elephants, which have ample food throughout the year, have small home ranges, while the elephants in the Namib Desert have huge home ranges and can cover between 25 and 70 kilometers per day. In the African savannah or the Asian monsoon climates, the food an elephant eats, and therefore the degree to which it moves around, depends on the time of year. Cynthia Moss writes:

> One of the things to emerge was the intriguing ability of elephants
> to change their behavior under various ecological conditions.
> The second thing to draw my attention was an overall change in
> migratory patterns of the elephants in the previous five years.[9]

This behavioral flexibility is related to the elephant's ability to find food even when it is scarce. In a year when rainfall had been very low and food resources scant, Moss observed how

> the elephants spread themselves throughout the park and surround
> ing areas, with each family returning to its clan's dry-season home

range and tending to move on its own. During the severe drought months of 1976 even some of the family units broke down, with a single female and one or two of her offspring forming a subunit.[10]

Elephant ecologist Philip Viljoen writes that

> during a recent five-year drought period more than 80 percent of the other 'desert' mammals like gemsbok and springbok died. As far as could be ascertained, not one of the desert-dwelling elephants died during this period.[11]

These examples illustrate the elephant's resilience, its ability to change with changing conditions.

Not only the elephant's use of the pliant and strong organ of the trunk, but also the way it applies its massive body and tusks, as well as its dexterous lips and tongue, reveal its flexible and unspecialized nature. In feeding, the elephant thrives on variety and at the same time can respond to inadequate or imbalanced food supplies by finding new sources of food and different areas to feed in, and by adjusting its social behavior. In its anatomy, physiology, and ecology, the elephant is a flexible creature.

Lifelong Change and Social Interactions

Elephants live remarkably long lives, up to about 65 years in the wild. This long life includes an equally untypical long period of growth and development. Most large mammals enter sexual maturity when they are a few years old and stop growing soon thereafter. Not the elephant.

Females enter puberty and begin mating at around 12 years of age. The gestation period is long, about 22 months, so that a female becomes a mother of a single calf at 14. (Occasionally twins are born.) She gives birth every four to eight years or so until she is in her fifties. Since the oldest females are the largest, they evidently continue to grow throughout life, although growth after 25 years is imperceptibly slow. Such lifelong body growth is not typical in mammals.

In males, adolescence is long and drawn out. They start producing sperm at around 14 years. At this time the two temporal glands, located between the eye and ear on each side of the head, begin to periodically secrete a fluid that smells like honey and even has some chemical resemblance to it.[12] These pubescent males don't begin mounting until a few years later and even then they usually back off when older males appear. Males go through a secondary growth spurt after puberty, so that a 19-year-old male becomes larger than an adult female.

In its early twenties, a male finally becomes sexually active and, except during mating, associates only with males. Full maturity in males is marked by the onset of musth between 25 and 30 years of age. "Musth" is the term used to describe the periods of heightened sexual activity and aggression in the adult male's life. During musth, foul-smelling fluids stream from the temporal gland, a stark contrast to the sweet-smelling secretions of younger years. These secretions are not only an expression of a physiological state, but are also important in social interactions. When young males smell an older male in musth, they tend to recede and avoid getting too close to him. Older males have regular musth cycles, and most probably remain sexually active until death. Males continue to grow, if ever slower, throughout their lives.

The extended period of physical development is one facet of an overall pattern of lifelong change that is also expressed in the rich and changing social relations an elephant experiences in the course of its life. This is especially true of females, who live in extended family groups their whole lives and are scarcely alone for a single minute of the day. In contrast, males become independent of family groups and tend to be more solitary the older they get.

A family group consists of grandmothers, mothers, sons and daughters, aunts, cousins, and nieces and nephews. A family group interacts periodically with other family groups, such meetings varying on a daily and seasonal basis. In this way an elephant is embedded in a relatively stable, but slowly changing familial environment woven into a larger context. As writer Charles Siebert puts it nicely, "A herd of them is, in essence, one incomprehensibly massive elephant: a somewhat loosely bound and yet intricately interconnected, tensile organism."[13]

With each stage of maturation an elephant's life changes. Young elephants always remain in close physical proximity to their mothers and can often be seen standing under the mother's body, between her legs. Often, just before the mother gives birth to a new calf, her behavior changes and she weans her four-year-old. While the connection to its mother remains

FIGURE 2.3. Mixed-age group of African elephants at a watering hole; Chobe River, Botswana.

preeminent, the weaned calf spends increasing amounts of time playing with other calves.

Young females seem to have an irresistible urge to be around, play with, and care for infants. Eight-year-old females will often be seen near their own or another mother, interacting with their younger siblings or cousins. With puberty, mating, and pregnancy, the female's life changes radically, but she is not suddenly an adult. As Cynthia Moss observed, a young, first-time mother appeared "upset and apparently confused about what to do" when her baby was born, while an older mother, who had given birth several times, appeared "relaxed and competent from the moment the baby was born."[14] The young mother receives help from other adult females, and sisters and nieces will begin looking after her calf, as she had done for other mothers a few years previously.

Her calf grows, and she will give birth again and again. All this time the female elephant matures behaviorally, her position within the family group gradually shifting. New interactions arise as her daughters give birth and her care extends to her grandchildren. At some point her mother will die, and she, or perhaps an older sister or cousin, will become the family matriarch.

Each family group has one such dominant female, which is usually between 40 and 60 years old. While continuing to gain experience, the matriarch still retains the flexibility to revert, when needed, to previous forms of behavior. So if one of her daughters or nieces dies, she once again becomes the protecting, care-giving mother. When a group led by an older matriarch comes into contact with other family groups, it discriminates clearly between little-known and better-known groups (indicated by bunching and smelling behavior).[15] In contrast to this, a group led by a younger matriarch does not make such distinctions between groups. Is this an elephantine version of the wisdom of old age?

Up until the 1989 ban of ivory trade, many older matriarchs in some regions of Africa were killed by poachers. This disrupted the social organization of family groups for up to 15 years.[16] The remaining females were more highly stressed and their reproductive rate slowed. Only when a female became older and new old-young female groups developed did the conditions normalize.

The importance of the experienced matriarch is underscored by observations comparing groups with a matriarch with those that—due to poaching—had no older matriarch.[17] In times of drought the family groups with the older matriarchs moved away from the river, where there was little to eat, and moved into areas where they could find more food. In contrast, the groups with younger females stayed near the river and suffered significantly greater calf mortality than in the groups with the experienced matriarchs.

Observations concerning "rogue" African elephant bulls show in a vivid way how essential interaction is between elephants of different age groups. For example, between 1992 and 1997 young elephant bulls in musth killed more than 40 white rhinoceros in a South African national park.[18] These bulls were orphans that had been introduced into the park. They were survivors of elephant kills ("culling operations") to reduce the number of elephants in another national park. The young males were less than 10 years old when they were introduced into the park. At this stage of development they would normally still be part of a family group and then, when older, would be associated with other younger, but also older bulls. All this contact was missing, and precisely these bulls carried out the abnormal rhinoceros killings.

These young males had musth periods that lasted much longer than normal, which is one sign of their overt aggressiveness. Park officials then introduced six older bulls into the park. The musth periods of the orphaned males—which had had no contact with older males up to that point—shortened significantly. The rhinoceros killings ceased. This remarkable change shows the power of inter-elephant contact. It is astounding that the physiological changes correlated with musth can be altered by the presence of another, older male. The younger males displayed abnormal behavior due to the lack of contact with a family group and older bulls, but this behavior could also be transformed when contact was normalized. The young bulls—they were around 20 years old—remained impressionable.

The elephant's long life is marked by lifelong change, physically and behaviorally. This means that an elephant also remains behaviorally flexible throughout its life. In mammals generally, behavioral plasticity is most pronounced in childhood, before sexual maturity and the cessation of bodily growth. After sexual maturity, behavior becomes more rigid, which is why animal trainers work with very young animals, when their behavior is still more open to outer influences.

One would expect that with their slow maturation elephants would retain a more open and flexible learning capacity for a longer period of time. And this is the case. In Asia, for example, it was a widespread practice *not* to train work elephants at too young an age. The elephants (of both sexes) were allowed their extended childhood and only *after* puberty did training slowly proceed.[19] Only at about the age of 20 has the Asian elephant matured enough to become a full-working animal. It is also possible to capture adult elephants and train them, something one never does with other large mammals.

FIGURE 2.4. Young bull, with ears spread, makes a mock attack after having loudly snapped off a tree branch; African elephant, Moremi Wildlife Reserve, Botswana

We associate playful behavior primarily with the behavior of young mammals, but as M. Krishnan writes:

> Elephants are among the few animals that play even as adults. When bathing in company in a forest pool, it is not only the juveniles that revel in play: even the adults bump into, push down, and roll over one another with abandon—perhaps they find the sudden lifting of their ponderous weight off their feet by the water exhilarating. On land too, adult bulls may indulge in a long bout of play with their trunks, not in a tug-of-war so much as a pushing match, or in chasing one another.[20]

In the context of lifelong flexibility we must not forget the trunk. As we have seen, the trunk is an unspecialized organ that allows the elephant to explore and interact with its environment in the most adaptable and diverse ways. Putting it simply, having a trunk, the elephant can't stop being flexible, even after sexual maturity. The elephant's overall openness and adaptability in behavior throughout its long life expresses itself at any given moment in the activity of the trunk. The trunk, like the human hand, makes visible what open, explorative learning behavior is all about.

Teeth Reveal the Whole

Being the hardest, most crystalline part of the animal body, teeth are well-defined structures, and their size, shape, number, and position are characteristic for each species of mammal. If you ask zoologists or paleontologists what small part of the body they would like to have in order to identify an animal, it would be a tooth.

By all accounts, elephants have strange teeth. Let's begin with the tusks, which are modified teeth. All elephants are born with small deciduous ("milk") tusks that fall out after approximately one year. The permanent tusks then erupt. The tusks are deeply rooted in the upper jaw, with about one-third of the tooth being hidden in the jawbone (see Figure 2.12). This part of the tusk has an inner pulp cavity from which growth originates.

The tusks grow throughout the elephant's life, becoming longer, thicker at the base, and ever more deeply rooted in the upper jaw. (Outward growth is balanced by inward anchoring.) Old African elephant bulls have the largest tusks, which can be over 10 feet long, weighing

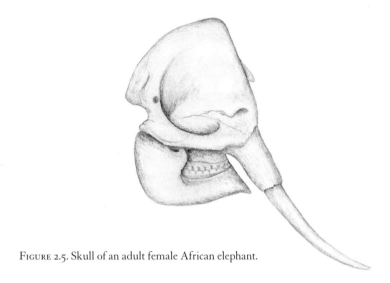

Figure 2.5. Skull of an adult female African elephant.

more than 200 pounds. Only in the female Asian elephant do the tusks remain, as a rule, small.

In general, the tusks grow downward and then curve upward. An individual elephant can often be identified by the unique curvature of its tusks. The tusks are used to dig, carry, and cut through tree bark. They are also wielded as gouging weapons. An elephant often employs one tusk more than the other, and this tusk then grows thicker and heavier, while the tip becomes more blunted.[21] The extra growth is evidently induced by the stresses and strains met in using this "appendage."

As the tusks grow, so does the skull, becoming higher and wider. The neck muscles holding the ever-weightier head continue to enlarge, as do the spines of the thoracic vertebrae to which these muscles attach. This increased weight is, in turn, supported by the growing leg muscles and thickening leg bones. We see how the growth of a single feature is related to the whole body. The lifelong growth of the body described in the previous section now shows its concrete relation to the growth of the tusks.[22] But that is not all.

Because of their position in the front of the upper jaw (in the premaxillary bone), the tusks are considered to be incisors, although they have the conical form of canine teeth. The elephant has no lower incisors and no canine teeth at all. The remaining teeth are molars in the rear of the jaw. Just as the elephant's incisors are atypical for mammals, so are its molars.

When an elephant is born, it has two to three molars in each side of the lower and upper jaws. In the course of time these teeth wear and move forward in the jaws. The roots then begin to be reabsorbed and pieces of the teeth break off. As one molar is being lost in the front, a new one begins to erupt from the back of the jaw. Each new molar that comes in is larger than its predecessor (see Figures 2.6 and 2.7). This means that the jaws also grow, mirroring the continuous growth of the tusks and with them influencing the growth and transformation of the rest of the head and body.

During the course of the elephant's life, a total of six molars pass through each side of the upper and lower jaws. At any given time an elephant has parts of one or two molars (one coming in from the rear and one being worn down and breaking off at the front) in each half of the upper and lower jaws. Natural death in old age is connected with the wearing and loss of the sixth and last molar. When an elephant is between 40 and 50 years old, the fifth molar is lost and the large sixth molar, which has been emerging over a decade, is the only tooth remaining in each half of the upper and lower jaws. Over the next 15 to 20 years it wears to the

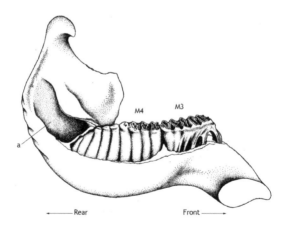

FIGURE 2.6. Left half of lower jaw (mandible) of the African elephant; medial view. The third molar (M_3), partially broken off and strongly worn, is being replaced by the fourth molar (M_4). Behind M_4 is the space (a) in which the fifth molar is developing. (Adapted from Hanks, 1979)

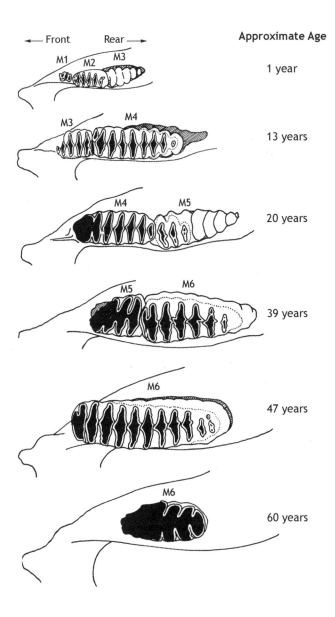

FIGURE 2.7. Tooth development in the elephant. This figure shows the right half of the lower jaw of the African elephant, viewed from above. Each of the six molars (M_1 to M_6) forms in the rear part of the jaw, gradually erupts and moves forward, wears due to usage, and then breaks off as it becomes fully worn at the front of the jaw. Note that the teeth become progressively larger. At approximately 60 years only part of the last molar still remains. (Modified after Laws 1966)

gums. Old elephants can be observed eating soft swamp plants, since they can no longer grind harder plant fibers. Death will soon follow.

Other animals die due to tooth wear, but in the elephant this is the last phase of a lifelong process of new tooth production and ensuing wear. Most mammals go through a change of teeth in the first few years of their lives. The permanent teeth are in place not long after sexual maturity and remain there until death. The phase of tooth change is therefore concomitant with rapid body growth and intense learning. The completion in the change of teeth marks the end of an animal's youth.

From this perspective, an elephant's youth extends through most of its life, just what we have seen from other points of view.[23] The physiological processes involved in tooth formation—the ever growing tusks and the prolonged formation of new molars—continue into old age. What typically comes to rest in the head of a mammal at an early age stays in process in the elephant. Isn't this a "tooth-appropriate" expression of flexibility, of staying-in-flux?

So the elephant's singular dentition now appears as an integral feature of its overall character, as expressed in the qualities of sustained change, learning, and flexibility.

Flexible Ideas

In trying to understand an organism, ever-new perspectives open up as one contemplates its characteristics, moving from one to the next. In this way we have been able to see that such seemingly disparate features as learning and tooth development are connected.

But not all features of the elephant are adequately illuminated through the qualities of flexibility and constant change. In fact, there is a danger of letting the richness of such qualities die into abstract concepts under which we then subsume all other phenomena. As Goethe put it, the human tendency to take "pleasure in a thing only insofar as we have an idea of it" can become tyrannical as "thought forcibly strives to unite all external objects."[24] Ideas then become "lethal generalities."[25]

We are never free from this problem. But if we take it seriously, we will strive to form our ideas in the course of intense immersion in the concrete phenomena. These ideas then have a proximity to the phenomena that makes them vibrant. Discovering flexibility in the trunk and then in behavior is very different from having preconceived notions like "animals are machines" or "animals are survival strategies" that do *not* arise out of a consideration of the animal itself, but are brought in as presuppositions. These latter may be powerful explanatory constructs, but, since they are imposed from without they always give a very limited and thereby skewed picture of the actual phenomena, especially when they are mistakenly understood as encompassing the *whole* animal.

Once we have gained an idea through a study, we can also illuminate new phenomena with it. The elephant's change of teeth appeared in a wholly new light after we had seen how in other respects it remains an animal in transition throughout its life. By returning again and again to the phenomena with an interest to discover *new* features, we can temper the tendency to fall in love with an already-formed idea and, in our passion, tyrannize the phenomena with it.

In this dynamic, our approach to phenomena and concepts becomes fluid. The more we can make our ideas like the trunk of an elephant, exploring an object and thereby taking on its form and taking in its qualities, the more what we reveal will be the object and not our own predilections.

The Sensitive Giant

An adult African elephant bull can weigh over 10,000 pounds and needs to eat around 400 pounds of food per day. The legs must be strong and stable to carry such an enormous weight. In contrast to the legs of other weighty (graviportal) mammals like hippos and rhinos, the elephant's legs are long and straight (see Figure 2.8). They actually resemble a human leg more than that of other large four-legged mammals (Figure 2.13). The bones are embedded in thick layers of muscle, giving the legs their massive rounded, columnar appearance.

The bones of the legs are themselves exceedingly heavy and have no marrow cavity—a long, tubular space normally running the length of the shaft of long bones in mammals. The larger a mammal, the greater is the portion of its total body weight made up by the skeleton. An elephant's skeleton makes up about 25 percent of its total body weight, whereas a lion's skeleton makes up only 13 percent of its total body weight.[26]

The elephant's relation to weight comes into view more clearly when we discover that even a fast-moving elephant will have at least one foot

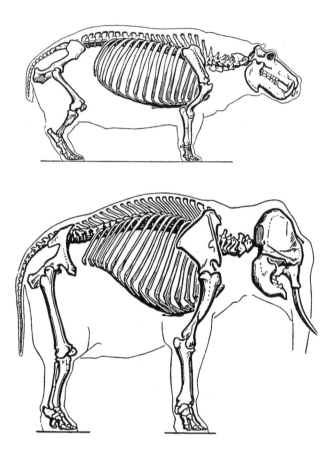

Figure 2.8. Skeletons of the hippopotamus and the Asian elephant. (Adapted from Tank 1984)

on the ground at all times. It never runs, which involves all four feet being in the air simultaneously for at least a short time, although when moving at full speed (up to 25 km per hour), the hips and hind limbs make motions similar to those other quadrupeds make when they run.[27] The elephant also cannot jump.

Though in this sense bound to the ground, the elephant does not strike one as being overly encumbered by the weight it carries. The long, pillar-like legs appear as sturdy supports for the raised voluminous body. Moreover, the way the elephant moves its bulk is special. In a detailed analysis of animal movement, Gambaryan describes "the smoothness of all movements" as one of the salient features of elephant locomotion.[28] (He is not thinking of the trunk in this context, although it is the epitome of smooth motion.) The body makes almost no vertical movements in walking. The massive frame seems to glide almost weightlessly forward, since no weighty up-and-down thrusting accompanies its motion.

The elephant can also, despite its bulk, move very quietly. While camping out in a wildlife preserve in Botswana, I was awakened in the night by a solitary elephant loudly breaking off tree limbs. I could follow its movement because of the cracking branches, not because I heard *it*. The moment the branch breaking stopped, I heard nothing and lost track of its movement. The next morning I discovered tracks just ten feet in front of the tent—the elephant had moved silently right in front of me and wandered off into another area.

An elephant's feet rest on the ground on large, roundish-to-oval surfaces and at first sight appear only to confirm the impression of rounded massiveness that characterizes the whole body. But when the elephant moves, we can see how carefully it can place its feet and how the pliant soles flow around any uneven hard surface that they tread upon.

This pliancy is related to the internal structure of the elephant's foot. Although one might not expect it, the elephant has five toes in the bony skeleton of its foot. The bones of the foot do not rest on the ground; instead, they are held at an angle and the tips of the toes carry the brunt of the animal's weight. Supporting the toes and foot is a cushion of fat embedded in elastic fibers (Figure 2.9). In walking, this cushion continually changes shape, compressing and broadening when the foot

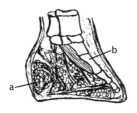

Figure 2.9. Diagram of the elephant's foot, longitudinal section. Left: foot lifted. Right: standing; when bearing the full weight of the body, the fat cushion (a) compresses. a: fat cushion; b: toe bones. (Redrawn after Gambaryan 1974)

bears weight, and rebounding to a more columnar form when the foot leaves the ground.

As elephant biologist Sylvia Sikes points out, "It is the possession of this shock-absorbing, internal cushion, and self-adjusting sole, that enables this enormous animal to walk over rough terrain and be inaudible to human ears."[29] This cushioning action is not merely a passive effect, since the muscles of the legs and the feet modulate every movement. The feet, especially the forefeet, have many muscles that allow extension, flexion, and lateral movements of the toes, giving them an unexpected internal mobility. In addition, the bones of the lower leg—the ulna and the radius in the forelimb and the tibia and the fibula in the hind limb— are "well developed and distinct bones, thus allowing the feet to achieve considerable rotation without loss of supportive strength."[30]

While most of the elephant's tactile exploration of the world occurs through its trunk, the feet are also used as tactile organs. Joyce Poole describes the outcome of throwing a rubber flip-flop sandal to a teenage male elephant:

> He first stabbed it with the tip of his tusk and then used it to scratch the underside of his trunk.... Finally he put it in his mouth and chewed it gently, turning it round and round slowly with his large tongue. After several minutes of such examination he tossed the shoe up in the air behind him.... He reversed several steps and reached out to touch it gently with his hind foot. [He] touched it carefully from all angles with both hind feet.[31]

This example illustrates beautifully how the elephant immerses itself in tactile experience, all the way from trunk and tongue to the soles of its immense feet.

This same elephant could, in the next moment, tear off a large branch with its trunk, and then use both the trunk and the feet to position, step on, and break the branch, enabling it to tear off pieces of bark to eat. Both trunk and feet are strong, yet sensitive and agile. These contrasting abilities deeply impress everyone who observes elephant behavior.

As Sylvia Sikes writes,

> The rapid alternation between movements of fastidious delicacy with which tiny berries and buds are selected and eaten, and

movements requiring tremendous brawn that send gigantic trees crashing to the earth, is always astounding.[32]

The Head in the Context of the Whole

A standing elephant makes a self-contained and calm impression. There are no small, nervous movements. While the trunk may be moving back and forth and the ears flapping, the head is held high and still. In fact, the elephant can hardly sink or raise its head at all, being the only four-legged mammal that in standing cannot lower its head to the ground. The elephant's neck is very short so that the back of the head nearly touches the shoulders. Externally, head and torso meld into each other, giving the elephant a compact appearance.

In other long-legged mammals (antelope, zebra), a long neck facilitates the movement of the head to the ground for feeding. At the same time the skull lengthens, with the jaws working as appendages to gather food (see Figure 2.10). By contrast, the elephant's neck is short, as is the

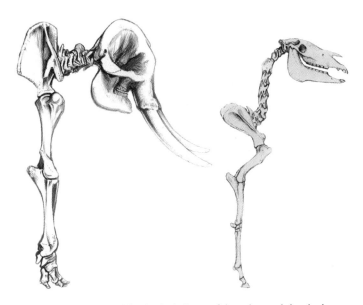

FIGURE 2.10. Head, neck, and forelimb skeleton of the zebra and the elephant (not to scale). Note the rounded head, very short jaw, and short neck of the elephant. In contrast, the zebra has a very long neck and a long skull. (Modified after Kingdon 1989)

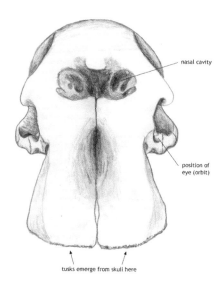

Figure 2.11. Frontal view of the skull of an African elephant without tusks. (Redrawn after Figure 5 in Van der Merwe et al. 1995)

skull, while the trunk lengthens as the organ used to gather food. This is a wonderful example of what is known in comparative morphology as compensation or the correlation of parts. One part of an animal does not develop just on its own; when one part changes the whole animal changes. The characteristics of an animal are finely interrelated.

The elephant has a remarkable head. Externally we notice this immediately in the unique appendage of the trunk, in the tusks, and in the large ears. With all these organs the elephant extends out into the environment and with trunk and ears takes the environment into itself.

The elephant's overall gestalt is characterized by two primary qualities: the compact body and the columnar verticality of the limbs. In both, rounded forms dominate. When we look at the skull, we find these qualities again (see also Figure 2.14). While the skull of most four-legged mammals is long (front-to-back) and narrow (side-to-side), the elephant's skull is high (top-to-bottom) and short (front-to-back). It is much more compact and rounded, mirroring the overall bodily form. This self-contained appearance of the skull is pierced by the downward curving and pointed tusks, and expanded in the large, often flapping ears. The lower jaw does not protrude forward and is tucked under the

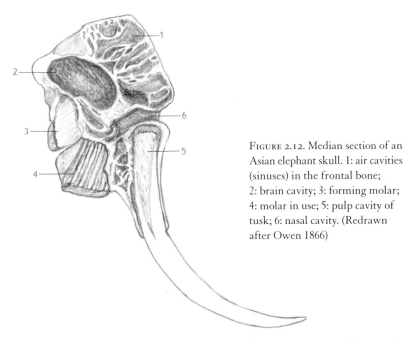

FIGURE 2.12. Median section of an Asian elephant skull. 1: air cavities (sinuses) in the frontal bone; 2: brain cavity; 3: forming molar; 4: molar in use; 5: pulp cavity of tusk; 6: nasal cavity. (Redrawn after Owen 1866)

high upper part of the skull. The parts of the elephant skull are ordered vertically, while in other four-legged mammals they are ordered behind one another.

The long snout of a zebra, extending from its long neck, functions as an arm. In the elephant this has been replaced, so to speak, with the mobile trunk, and as a result the whole skull forms differently. The trunk merges with the skull in proportionately very large and broad nasal cavities. Viewed from the front, the skull looks very odd (see Figure 2.11). The nasal cavities appear, at first glance, to be closely set eye sockets, whereas in reality the eye sockets are small and lower down on the sides of the skull. The nasal cavities, the bony and air-filled extension of the trunk, dominate the upper front part of the skull.

Behind and above the nasal cavities rises the high and pronounced forehead—another unique feature of the elephant skull. One would assume that this dome, which increases the overall rounded appearance of the elephant, houses the brain, but it does not. The brain cavity takes up only a comparatively small space at the rear of the skull. The forehead is, surprisingly, filled with an intricate network of air cavities, which is anatomically comparable to our frontal sinuses (see Figure 2.12). In the elephant these cavities have expanded to a remarkable degree. If we think of the air-filled trunk as a huge nose extending into the world, leading into the air-filled nasal cavities of the skull, which ultimately lead down to the proportionately large lungs, then we realize the intense relation the elephant has to air.[33] And now we find an air-filled vault resting above the nasal and respiratory part of the skull!

The sinuses of a newborn elephant are very small and develop gradually throughout life. Virtually every bone of the skull forms smaller or larger air-filled sinuses over time. The result is that, despite its immense size, the upper skull of the elephant is comparatively light. Sylvia Sikes noticed that when an elephant walks on the bottom of a lake or river, or swims, its head is held higher, apparently floating through its buoyancy.[34]

Like other parts of the elephant, the sinuses certainly have multiple functions, although biologists often have a strong urge to "explain" them in terms of one particular function. In addition to providing buoyancy in water, they create a large surface for muscles to attach to, and they may even secrete fluid when the elephant has no access to an external source of water. And, as we will see later, they play a role in transmitting sounds.

The Sensitive Boundary

Its deeply wrinkled, nearly hairless skin encloses the elephant's voluminous body. The elephant looks ancient from youth on. The skin is thick—over an inch on the rump—but it is not an inert covering. It is imbued with fine tactile responsiveness and mobility. An elephant can twitch the skin muscles back and forth to cause a fly resting on its skin to move away.

But the skin of the elephant is also sensitive and demands care. Pooling observations made by researchers on captive and wild elephants, Chevalier-Skolnikoff and Liska found that over 80 percent of tool-using behavior carried out by elephants, like scratching themselves with a stick held in the trunk, were in the context of body care.[35] If conditions allow, an elephant will bathe and wallow in mud daily, finishing its bath with a shower of mud, soil, or sand. This natural garment clothes the skin until the next bath or rain. As a result, elephants carry the hue of the soil of their region.

That the elephant covers its skin by means of its environment may not seem astonishing inasmuch as it lacks the protective fur of other animals. The elephant actively augments its sensitive boundary to the environment by its own activity. At the same time, in the activity of bathing and covering the skin, we can see the elephant's tendency toward tactile contact, which we saw also in the trunk and feet.

It is interesting that elephants show an additional form of "covering behavior," as naturalist Jonathan Kingdon calls it.[36]

Two descriptions:

> On one occasion I came upon the carcass of a young female who had been ill for many weeks. Just as I found her, the EB family, led by Echo, came into the same clearing. They stopped, became tense and very quiet, then nervously approached. They smelled and felt the carcass and began to kick at the ground around it, digging up the dirt and putting it on the body. A few others broke off branches and palm fronds and brought them back and placed them on the carcass.[37]

> I have known cases, where upon running away from an elephant, the people have fallen down and, expecting to be crushed at any moment, have found to their surprise that the elephant stops near to them and proceeds to cover them with mud and leaves.[38]

Elephants also cover up the remains of animals (lions, for example) and human beings they have killed. This kind of behavior is, to my knowledge, unique to elephants.

Dying and Death

Whether such covering behavior represents a form of burial is open to question. But it is evident that elephants have a special relation to dying and death, and to the dead remains of their fellows.

When a matriarch of a family was close to dying and falling to the ground, she was aided not only by members of her own family, but also by other elephants.[39] For example, a matriarch of a different family group lifted her with her tusks up to her feet. When the matriarch died soon thereafter, members of her own family and of other families came to the carcass, sniffing and touching it. Periodic visits by individual elephants continued over five days.

Once the body has decomposed and only the bones remain, elephants still show a selective interest in elephant bones in contrast to bones of other animals. It doesn't matter whether they are the bones of a relative or of some unrelated animal.[40]

Elephant researcher Joyce Poole vividly describes the relation of elephants to their dead:

> There is something eerie and deeply moving about the reaction of a group of elephants to the death of one of their own. It is their silence that is most unsettling. The only sound is the slow blowing of air out of their trunks as they investigate their dead companion. It's as if even the birds have stopped singing. Just as unsettling is the way elephants back into their dead. Although elephants use their front legs for killing, by kneeling on their victims, they have a way of walking backward and using their sensitive hind feet surprisingly delicately for waking up their babies and touching the dead. Using their toenails and the soles of the feet, they touch the body ever so gently, circling, hovering above, touching again, as if by doing so they are obtaining information that we, with our more limited senses, can never understand. Their movements are in slow motion, and then, in silence, they may cover the dead with leaves and branches. Elephants' last rites? A wake, a death watch, the calling up of the elephant spirits? Elephants perform the same rituals around elephant bones. They approach slowly and silently, and then the touching begins, slowly, as they deliberately, carefully turn a skull over and over with their trunks, touching, hovering over the long bones with their hind feet.[41]

Listening and Vocalizing

Through its fine sense of touch the elephant experiences its world in immediate bodily contact. It takes in a larger sphere of its environment through its discerning sense of smell. An elephant usually notices you with its sense of smell and not with its eyes. It raises its trunk, sways the tip back and forth, and then comes to orient it in your direction. Close your eyes and wave your hand slowly through the air, imagining that your fingers are smell sensors, and you can get an inkling of what it might be like when an elephant's undulating trunk embraces the scents wafting in the air.

The elephant has another sense that allows it to spread out over still larger spaces—its sense of hearing. The large size of its ears (pinnae) already indicates the prominence of hearing in elephants. The moving surfaces help elephants locate sounds and direct them to the inner ear. (As with so many features of the elephant, the external ears are multifunctional. Large amounts of blood flow through them, and the blood cools down in the process, so the earflaps are also organs of temperature regulation.) The elephant also has unusually large and heavy ossicles (hammer, anvil, and stirrup) in the middle ear that conduct vibrations to the inner ear. Interestingly, elephants have a muscle that allows them to close the outer ear canal—another sign of flexibility in an unexpected place.[42]

Scientists have discovered that elephants hear and emit very deep tones inaudible to the human ear. Katherine Payne describes her discovery of the phenomenon of low frequency sounds and hearing in the Portland Zoo:

> While observing three Asian elephant mothers and their new calves,
> I repeatedly noticed a palpable throbbing in the air like distant
> thunder, yet all around me was silent.[43]

Since then extensive observations and experiments in Africa have shown that, depending on the atmospheric conditions — which vary with the time of day and season — elephants can communicate with each other via low-frequency sounds over distances of several kilometers. Although groups of elephants may lose visual contact, they don't lose auditory contact. When a female is in estrous, for example, she makes deep rumbling sounds, attracting males that are dispersed in the area.

Low-frequency sound communication makes understandable a phenomenon that had long puzzled elephant researchers. Viewing groups of elephants from the air, which were clearly not in olfactory or visual range of each other, researchers observed how the groups seemed to make coordinated movements and, for example, all make for the same watering hole. While listening, the elephants "hold perfectly still, raising and stiffening their ears and slowly swinging the head from left to right as if to localize the source of a call."[44]

Scientists have discovered that these low-frequency sounds, as well as reverberations from elephants stomping their feet during a mock attack, travel through the ground for even larger distances than through the air.[45] It may be that these massive creatures are also using the earth's mass to communicate over large distances. Not only may the large middle ear ossicles play a role in mediating these vibrations, but also "the elephant's body with a massive skeleton and pillar-like bones might be suitable for conducting the surface waves to the inner ear."[46] It's impressive to think of the elephant's entire body as an organ of hearing.

The deep low-frequency sounds that travel long distances through the air can only be produced and emitted by a large animal whose body acts as a voluminous resonator and transmitter for the spreading vibrations. (Think of a bass compared to a violin.) As in hearing, so also in the production of sound: the whole animal is involved. The large air-filled sinuses in the head can be viewed as part of the elephant's resonating body when it emits sounds.

When we try to imagine the elephant in its world, we first focus on the trunk as the center of its tactile and olfactory experience. Through these senses the elephant actively explores and takes in the qualities of the world it lives in. Through its large and uniquely framed body and its ability to hear low-frequency sounds, the elephant in the savannah lives in a sea of deep tones and vibrations, its boundaries stretching over kilometers. This large animal embraces through its rumblings a large environment.

But low-frequency sounds do not encompass the full range of elephant hearing and vocalization. In her book *Elephant Memories*, Cynthia Moss describes the impressive array of sounds made by elephants: bellowing, clicking of tusks, cracking of ears, groaning, growling, humming, moaning, rasping (of ears), rumbling, screaming, sluffing (of feet), snoring, squealing, and trumpeting. Virtually all of these sounds are related to communication between elephants; they are contact calls, alarm calls, and so on.[47]

The variety of sounds an elephant can make has a correlate in its fine ability to distinguish between different sounds. Elephant trainers make use of this capacity, and a fully trained work elephant in Asia can respond to dozens of whispered commands from its mahout. And elephants can even imitate sounds.[48]

As we saw in the manual flexibility of the trunk and in the variety of diet and habitat, so now we find in the elephant's hearing and vocalizations a remarkable array of possibilities. This spectrum extends from the outspreading rumblings resonating in the large and voluminous body to the close-up fine receptivity for whispered tones, revealing from one more facet of how the elephant unites contrasting features: voluminous bulk and sensitive discrimination.

Comparison with the Human Being

The last thing you would think of when you begin to study the elephant is to find similarities to the human being. But once you look beyond the glaring external differences and begin to build a picture of the elephant's various characteristics, you can't overlook the numerous features that elephants and human beings have in common. Various authors have been struck by such commonalities.[49] Wildlife biologist Douglas Chadwick summarizes:

> In fact, humans and elephants are together at the extreme end of the scale in terms of the number of years during which offspring are carefully tended by their parents. Both have young that mature only in their early teens, and both continue to care for them until that time. This derives from another basic shared quality: we are both unusually long-lived as mammals go. Such lengthy nurturing also presupposes a good deal of intelligence
>
> …. In terms of learning abilities, we and elephants are once more together at the extreme end of the scale…. The majority of what [other] animals need to know to survive is already built in, largely instinctual
>
> …. Like humans, elephants are designed to learn most of what they need to know. The extended period of nurturing is part of that process, and they continue learning throughout their long lives.[50]

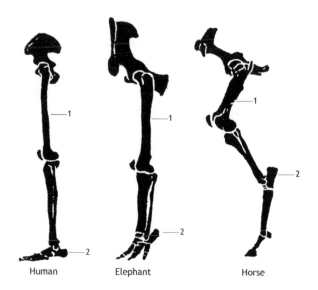

Human   Elephant   Horse

Figure 2.13. Leg of the human and hind limbs of the elephant and horse. Both the human leg and the elephant hind limb are very straight, and the proximal (body-near) bones are the longest, while the bones of the feet are short. In contrast, the more angular leg of the horse has very long bones in the feet (the heel bone [2] is high off the ground) while the proximal leg bones are proportionately short. 1: femur (thigh bone); 2: calcaneus (heel bone). (Adapted from Carrington 1959)

Related to all these qualities is the fact that both elephant and human being have flexible prehensile organs—the trunk and the hands. (But how different it must be to have only one such limb, which extends out of the head and at the same time has the ability to smell!) The freedom of movement these organs enjoy is made possible by the weight-bearing legs, which elevate and support the rest of the body. We have already seen that the elephant's straight legs resemble in overall structure those of the human being more than they do those of most four-legged mammals. But we should not forget that the elephant has four such columns, and not just two, to give stable support to its massive frame (see Figure 2.13). Interestingly, the two mammary glands of the female elephant are situated between the forelegs, that is, in the same position as in humans. This is unique among four-legged, non-arboreal mammals.

The elephant carries its head high off the ground and cannot reach the ground. Like our hands, the trunk functions as a limb to bring food

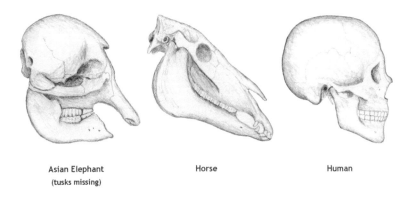

Asian Elephant   Horse   Human
(tusks missing)

FIGURE 2.14. Comparison of elephant, horse and human skulls; not drawn to scale.

and water to the mouth. As we have seen, the elephant does not have the horse's long bony snout; this front part of the skull is short, and what extends to the fore is the muscular trunk (see Figure 2.14). Not only is the elephant's skull short, but it also has a high forehead, and the relatively small lower jaw (mandible) is tucked under the cranium. The upright orientation of the skull is accentuated by the nearly vertical premaxillary bone, which carries the tusks. The skull as a whole is large in relation to the rest of the body. All of these features resemble a human skull more than they do a typical four-legged mammal's. In both legs and skull the elephant shows a tendency toward verticality, a tendency that dominates human morphology.

It would take a whole other chapter to show how everything in the human body is configured according to the upright posture and to show its relation to the psychological and spiritual nature of being human. Let one quote suffice. Johann G. von Herder, a friend and colleague of Goethe's, had a keen sense for how uprightness is a holistic quality that infuses all aspects of human life. He wrote:

> Because the human being has to learn all things, because it is our instinct and calling to learn everything like our upright gait, we learn to walk by falling and come often to truth only through error. The animal is carried forward securely in its four-legged gait; the more strongly expressed proportions of its senses and drives are its guides. The human being has the advantage of a king to look to far

horizons, upright and with head held high. Of course, we also see much darkly and false. We forget our steps, only to be reminded when stumbling on what a narrow basis the whole head- and heart-edifice of our concepts and judgments rests…. The human being is the first to be set free in creation. We stand upright. The balance of good and evil, of false and true hangs in us. We can search, we shall choose. Just as nature gave us an over-viewing eye to guide our gait, so also do we have the power, not only to place the weights, but— if I may put it this way—*to be the weights* on the balance.[51]

Herder captures in a beautiful and concise way the gifts and predicaments of being human—the ability to gain distance from the web of natural workings and to make free decisions using our own capacity of judgment. But we also err and wreak havoc in the world when we act out of ignorance. We must find our balance, our own moral relation to the world, just as we continually balance our body on the "narrow basis" of our two feet.

The elephant, in contrast, is embedded in its world in a harmonious way. It has the gift of being an animal. And yet, with its long columnar legs, its shortened and upright head, its flexible trunk to interact flexibly with the world, its long development and lifelong learning capacity, it has much in common with us, while remaining more embedded in a natural ecology. As outwardly dissimilar as elephant and human being may first appear, the more we get to know the elephant, the more we get to know—if not in an evolutionary sense—one of our closest relatives on earth.

Elephantine Intelligence

The elephant is well known for its intelligent behavior. Let's look at various examples of non-trained elephant behavior that the people making the observations considered intelligent:

> If he cannot reach some part of his body that itches with his trunk, he doesn't always rub it against a tree: he may pick up a long stick and give himself a good scratch with that instead. If one stick isn't long enough he will look for one that is.[52]

> On many occasions I have watched an elephant pick up a stick in its trunk and use it to remove a tick from between its forelegs.

I have also seen elephants pick up a palm frond or similar piece of vegetation and use it as a fly swatter to reach a part of the body that the trunk cannot.[53]

If he pulls up some grass and it comes up by the roots with a lump of earth, he will smack it against his foot until all the earth is shaken off, or if water is handy he will wash it clean before putting it into his mouth.[54]

Elephants have picked up objects in their environments and thrown them directly at me, undertrunk, with surprising, sometimes painful, accuracy. These projectiles have included large stones, sticks, a Kodak film box, my own sandal, and a wildebeest bone…. Elephants have been known to intentionally throw things at each other in the same circumstances: during escalated fights and during play. Elephants have been known to intentionally throw or drop large rocks and logs on the live wires of electric fences, either breaking the wire or loosening it such that it makes contact with the earth wire, thus shorting out the fence.[55]

[In India an] elephant was following a truck and, upon command, was pulling logs out of it to place in pre-dug holes in preparation for a ceremony. The elephant continued to follow his master's commands until they reached one hole where the elephant would not lower the log into the hole but held it in mid-air above the hole. When the mahout approached the hole to investigate, he found a dog sleeping at the bottom; only after chasing the dog away would the elephant lower the post into the hole.[56]

[In South Africa] it was observed that an elephant, after digging a hole and drinking water, stripped bark from a nearby tree, chewed it into a large ball, plugged the hole, and covered it with sand. Later he removed the sand, unplugged the hole, and had water to drink.[57]

Many young elephants develop the naughty habit of plugging up the wooden bell they wear around their necks with good stodgy mud or clay so that the clappers cannot ring, in order to steal silently into a grove of cultivated bananas at night. There they will have a whale of a time quietly stuffing, eating not only the bunches of bananas but the leaves and indeed the whole tree as well, and they will do this just beside the hut occupied by the owner of the grove, without waking him or any of his family.[58]

As we can see from these examples, intelligent behavior allows the animal to deal with a concrete situation in a flexible and nonschematic manner. Or as Shoshani and Eisenberg put it, intelligence is "the capacity to meet new and unforeseen situations by rapid and effective adjustment of behavior."[59] Intelligence presupposes an ever-present ability to learn. Not unexpectedly, many of these examples show that the elephant's intelligence often manifests through the activity of its trunk: breaking off sticks that are then handled as an extended limb for scratching or swatting; throwing with the trunk; stuffing a bell with the trunk.

With such a flexible and dexterous prehensile organ, how could an elephant not be intelligent?

At the same time, these activities involve the whole animal in the coordinated use of different body parts and senses: sight and trunk are used in throwing, while foot and trunk coordination allows cleaning clumps of grass. Raman Sukumar, who studied the Asian elephant in India, describes a scene that clearly illustrates the elephant's complex behavior:

> Vinay [a solitary adult Asian bull] poked at the *bendai* tree with his left tusk, thrusting it up into the gash and splitting the bark. He grasped a portion with his trunk and tugged expertly with an upward flick, tearing off a four metre long strip. Another tug and the strip broke loose from the tree-trunk and came down. Vinay now began eating the bark, skillfully using his forefeet and trunk to break off small strips before transferring them to his mouth.
>
> After feeding for some ten minutes, Vinay did something that only an elephant can do so effortlessly. He turned towards the tree, and using his forehead and trunk, pushed the tree over. In a minute or so the tree was cleanly uprooted. Vinay tore just one more strip of bark from the tree and then turned away. Almost nonchalantly he began to pluck green grass that sprouted profusely from among burnt clumps. As he wrapped his trunk around a clump and pulled, the tender leaves came off quite easily from their dry bases. Stuffing one trunkful after another into his mouth, Vinay ambled along at a gentle pace.[60]

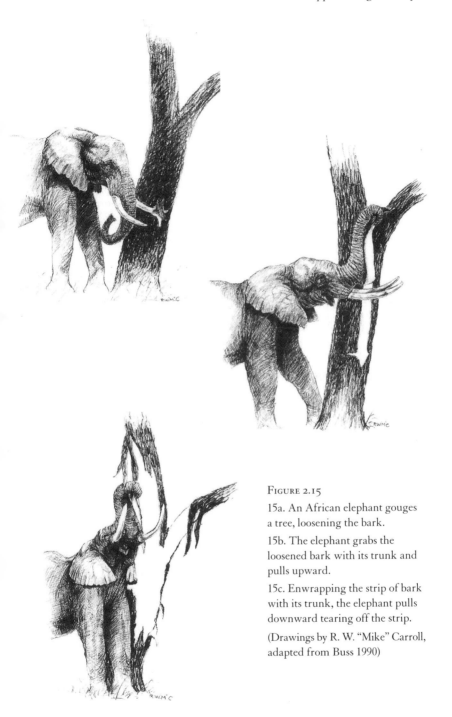

FIGURE 2.15

15a. An African elephant gouges a tree, loosening the bark.

15b. The elephant grabs the loosened bark with its trunk and pulls upward.

15c. Enwrapping the strip of bark with its trunk, the elephant pulls downward tearing off the strip.

(Drawings by R. W. "Mike" Carroll, adapted from Buss 1990)

The elephant's behavior flows from one activity to the next, engaging its brawn and dexterity as needed. The key to such actions and their sequence is that they are not automatic and prescribed. Intelligent behavior embodies plasticity—flexible interaction with experience. The elephant cleans off the dirt by smacking the clump of grass against its foot, but if it also perceives water nearby, it can then take the clump and submerge it in water to clean it further. It does not have just one "built-in" way to carry out tasks.

All of the above examples reveal what we would call purposive behavior. We have to be very careful, however, not to anthropomorphize an animal's behavior. We would clearly be anthropomorphizing if we imagine an elephant planning ahead of time to steal bananas, thinking over options, and then coming up with the idea of plugging the noisy bell. That is just putting a human mind in elephant skin. We also need to be careful in the case of the elephant that did not put the log on the dog; we should not immediately assume that the elephant took pity on the dog or had a conscious awareness it was about to kill it. This caution does not detract from the impressive act itself. Rather, it leaves us more open. We erase the possibility of understanding the elephant's unique kind of intelligence if we too easily transfer our experience to it. When we stay close to the perceived situation and restrain judgment, the unique and fascinating qualities of the animal become *more* vivid than if we make life too comfortable for ourselves and imagine it seeing the world through our eyes. After all, we want to see more than ourselves in the animal.

The scientist Herbert Haug carried out a detailed comparative study of the anatomy of elephant, dolphin (pilot whale in this case), and human brains to see if he could find out how the brains might relate to the intelligent behavior of these creatures.[61] The brains differ distinctly from one another, but all are large (see Figure 2.16). The elephant has the largest brain of all land animals; an adult elephant's brain weighs on average between nine and twelve pounds.[62] But, of course, the elephant also has the largest body of all land animals. The elephant's brain makes up about 0.08 percent of the total body weight. The human brain weighs three to four pounds and is also relatively large, making up 2 percent of our body weight.[63]

The brains of elephant, dolphin, and the human being are all highly convoluted, a characteristic that increases the surface area. These brains

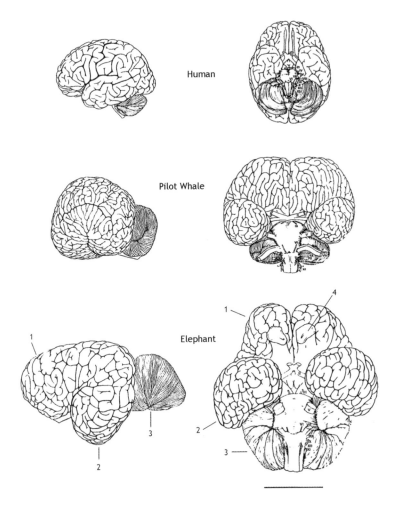

FIGURE 2.16. Brains of human being, pilot whale, and elephant, viewed from the side and from below. 1: frontal lobe cerebrum; 2 temporal lobe of cerebrum; 3: cerebellum; 4: olfactory nerves leading to the trunk, which are strongly developed in the elephant. Scale bar = 10cm). (Adapted from Haug 1970)

exemplify the well-known correlation between the degree of brain folding and the degree of intelligent, flexible behavior found in mammals.

But what is specifically elephantine about the elephant's brain? Three areas of the brain are noticeably enlarged relative to the other parts of the brain: the olfactory lobe, the cerebellum, and the temporal lobe of the cerebrum. Enlargement of part of the brain usually means that there are more neurons in that part. These neurons are connected to other parts of the brain and to the rest of the body via nerve fibers. The enlargement of the olfactory lobe is clearly connected to the fine innervation of the sense of smell in the trunk. The cerebellum has been found to be related to muscle coordination in other, better researched mammals. Since the nerve pathways in the elephant are not that well known, Haug can only make the clearly reasonable suggestion that the cerebellum's high degree of development is related to the highly coordinated trunk and limb movements. It is not surprising that the elephant's trunk-imbued intelligence, the focus of so many of its activities, is mirrored in the enlargement of parts of the brain connected to the trunk.

Why the temporal lobes are so large (proportionately larger than in any other mammal) remains a riddle. The temporal lobes are, among other things, related to hearing in mammals (and speech in the human being), so it does not seem too far off to conjecture that the elephant's ability to distinguish and communicate through a variety of sounds (including infrasound) may well be connected to the differentiation of the temporal lobes.

Haug's study led him to be skeptical about any claims that correlate intelligence and the brain too closely:

> From a qualitative point of view, the human being does not possess—compared to elephants and dolphins—a particularly high grade of cerebral differentiation that would provide the morphological basis for such a great difference in intelligence as is actually present.... The question must be asked, whether brain differentiation must necessarily be equated with human productive intelligence.[64]

There is a strong tendency in our times to want to localize intelligence—and other capacities—in the brain. It is a very unorganismic way of viewing that leads us to seek for a "command center" in the brain. Intelligence resides just as little (or just as much) in the brain as it resides in the elephant's trunk. It would be just as correct (or incorrect) to say that the elephant has its center of intelligence in the trunk as it would be to say that it is in the brain. If the elephant's trunk becomes lame, some of its intelligent behavior will be missing—just as, if part

FIGURE 2.17. African elephant bull. (After front jacket photo, Shoshani 1992)

of its brain is dysfunctional, some intelligent behavior may also not be possible.

In either case it could compensate for such injuries to a certain degree by engaging other body parts and brain parts. Intelligence resides everywhere and nowhere. Perhaps it is best to say we discover it in the intelligent activity itself, which is carried out and made possible by the *whole* animal. And in the elephant this whole is most vividly displayed in the use of the trunk.

Summarizing Picture

The massive and voluminous elephant stands firmly in the world, carried by its long pillar-like legs. The short, high-browed head connects to the torso with a very short neck that is hardly visible behind the large ears. This compact appearance is accentuated by the nearly hairless, gray and fissured skin. Internally, the elephant's weighty being finds expression in the high density of its limb bones and in the continuous production of dentine and enamel in the tusks and molars.

Not only is the head of the elephant uniquely shaped for a four-legged mammal, but from it emerge the most characteristic elephantine organs: trunk, tusks, and large ears. These organs reinforce the elephant's enormity. When you view from the front an adult African elephant with its ears spread, the animal appears as a massive wall, as broad as it is tall. But trunk, tusks, and ears also have a very different character from the rounded, self-contained head and torso. They radiate outward as organs of activity and expression. While the tusks protrude in rigid radiance, the trunk and ears move almost continually. Their large sweeping motions, which lack any hint of nervous haste, heighten the elephant's grandeur.

Interestingly, the center of glandular expression in the elephant—the temporal gland—is also located in the head. Through its secretions, the elephant expresses something about its age and current state that is finely perceived by other elephants. We also find a "headwards" movement of certain physiological functions and organs. While other four-legged mammals have the mammary glands between the hind legs, the elephant has one pair of mammary glands between the forelegs. The position of the vagina has also shifted forward—it does not lie directly below the anus, but has moved down and forward, being situated between the hind legs. It seems that the elephant's center of gravity, physically as well as functionally, has shifted headwards.

The elephant lives through trunk and skin at its body's boundary in intensive tactile contact with its surroundings. Watching a herd of elephants you see continual interaction—caressing, pushing, slapping, probing. And through its keen senses of smell and hearing, and its trumpeting calls and infrasound rumbles, the elephant enters a truly expansive world.

The trunk unites power and agility in singular fashion. We find this unity of largeness and delicateness, of enormity and sensitivity, in modified ways in nearly all elephant characteristics. With its finely modulating feet, a soft-treading elephant has little trouble moving silently through a forest, but can, in another moment, crash through the forest, bowling over

FIGURE 2.18.
Nursing infant
African elephant;
Chobe River,
Botswana.

trees or crushing a lion under its foot. The thick, leathery skin that appears so tough is also extremely sensitive, warranting continual care. The large, constantly moving ears are ideal for taking in and locating tones coming from afar, but can also hear the quietest tones and distinguish between subtle modulations. The elephant's unified being speaks through contrasts.

There is no more physically flexible organ in the animal kingdom than the elephant's trunk. While the trunk is clearly the elephant's focal instrument for living out its flexible nature, this paramount elephantine feature in fact expresses itself in the whole animal—physically, physiologically, and behaviorally. The elephant does not have to eat food of one type, it can shift from one food source to another; when given the opportunity it goes for variety. The elephant can live in different types of habitats—from the climatically uniform and food-rich rainforest to the extremes and dearth of the desert. But most elephants live in the more rhythmically changing savannah and monsoon climates, where they move with the changing seasons and the changing sources of food the seasons bring.

An elephant changes throughout its long life. Its primary growth phase lasts around two decades, but it continues to grow slowly until death. The tusks grow throughout its life, and, like no other mammal, the elephant's change of teeth in the molars never stops. There is an ongoing development of the new and discarding of the old, continuous physiological renewal. At the behavioral level we find this characteristic mirrored in the elephant's pronounced and lifelong learning ability. At any moment the elephant can adjust to new situations with its own unique form of intelligence.

In their family groups, elephants have intense contact and learn from one another. They go through a long phase of maturation and then develop different relations within the group as they grow older. The life of a young mother is very different from that of an older female or of the group matriarch. Or think of the orphaned aggressive young bulls that altered their behavior soon after older bulls entered their home range. In the elephant, the ability to change never ceases.

How Not to Lose the Elephant for All its Parts

You have now participated in my attempt to portray the elephant as a unique and unified being. I'd like to highlight here some of the features of my method. "Method" is a horribly dry term to express what I do to try to gain deeper glimpses into the way of being of an animal. What do I attend to and what do I try to avoid? (I'll say more about my approach throughout the book and summarize it in the concluding chapter.)

Apprehending wholeness demands a particular kind of attention and inner activity. First, when I have come to a certain understanding of some detail, instead of just progressing further in analysis I make myself—which is not easy—step back and ask, "How does this relate to the whole?" I may not yet have an answer, but by trying to place every detail into the larger context, I make sure I am not losing sight of the animal in all its parts.

Second, I try to withstand "explaining," by which I mean, in this context, finding a surmised single cause of the phenomenon I'm looking at. For example, when I look at the fact that elephants have very long straight legs I may be tempted to "explain" this fact by saying it is an efficient construction that enables such a large animal to bear its weight with minimal muscular exertion. In Darwinian terms, this characteristic gives the elephant greater survival chances. But since I also know

FIGURE 2.19. An example of cooperative elephant behavior. The elephant on the right had just used its bulk and a sideward thrust to push the elephant on the left up and out of the mud in which it was stuck; Chobe River, Botswana.

that the weighty hippopotamus has short angular legs, and survives in its way just fine, the "explanation" loses its compelling force. And this happens with virtually all single-cause explanations I have encountered. They just don't work. As far as I can tell, every biological fact has multiple relations that illuminate its function or form. Since single-cause explanations are usually false and tend to fix the mind in narrow pathways, I don't look for such explanations. When I find authors using them, I discard the explanation and let the phenomena stand on their own.

So with the iron will to always return to the whole and the discipline to hold back from shortcut explanations, I keep the path to the unity of the organism open. But to grasp this unity a third kind of activity is necessary. When I am studying a given phenomenon or reading what others have discovered, I make as vivid a mental picture as possible. I picture the exact form of the limb bones and how they articulate; I picture how the elephant feeds; I picture the formation of the teeth. Or when I have experienced a young bull making a mock attack and am later writing down my observations, I make sure I build up anew a vivid picture of what I have seen. By utilizing imagination and staying as close to the phenomena as possible, I try to create an exact picture. The goal is to achieve saturated inner images of the elephant's characteristics.

The remarkable thing is that when one builds these exact pictures over and over, moving from one characteristic to the next, patterns emerge. You begin to recognize how the characteristics form a whole—the unity begins to reveal itself. When you go back to characteristics you have studied before, they may suddenly express the unity you have discovered through another part. For example, over time I began to see that the elephant not only has a flexible trunk, but that flexibility is part of its whole being—flexibility in feeding behavior, flexibility in social interactions, flexibility in learning overall, and even flexibility rooted in the anatomy of its immense feet. Characteristics such as the long phase of growth, the long period of social maturation, the virtually lifelong change of teeth, and the lifelong learning capacity no longer appear as separate traits, but as expressions of a unitary being. When I have begun to grasp the elephant in this way as an interconnected whole, I can then set about describing and characterizing it. My goal is, in Goethe's words, to "portray, rather than explain."[65]

For this reason this account is not like typical biological descriptions of animals that depict the different anatomical and physiological systems of the body (skeleton, muscles, digestion, circulation, reproduction, etc.), followed, say, by a consideration of ecology and behavior. Instead, it is arranged according to themes, sometimes taking a particular part of the body or function as the starting point, through which I try, from a somewhat different perspective in each section, to express something of the elephant's unique character. It's a bit like hiking through a landscape and getting to know it from different vantage points. Every perspective takes in the whole, but does so by highlighting different aspects of it. In this way our own understanding is deepened. Through this activity—if it is successful—we can build up a picture of the elephant that expresses the unity of its different features.

Recognizing how every part of an animal manifests an underlying unity is an exhilarating experience. What seemed separate comes together, and we sense that we are seeing the elephant truly for the first time. We have a nascent answer to the question, Who are you, elephant? Making even the smallest steps in this direction opens up a wholly new appreciation and understanding of the animal.

On this path of discovery, the wisdom of nature becomes more tangible. In getting to know this wisdom, we find that our feeling of responsibility toward our fellow creatures on earth grows. We come to know ourselves as part of a world that is much greater than us. And when we learn from this world—when we glean something of its inner workings—we can also learn to care for it.

CHAPTER 3

How Does a Mole View the World?

I REMEMBER LONGING AS A CHILD TO EXPERIENCE—even just once—how an animal actually sees the world. To slip inside an ant and wander through the passages of the anthill, to see with the eyes of a squirrel. This longing, which I'm sure many of you know well, has not, in a straight forward way, been fulfilled. I can't get inside my cat—at least directly. And, if *I* were inside the cat, would I be seeing as me, or as the cat? It seems like an unsolvable problem—we can't get inside the animal.

Behaviorism in the twentieth century brought one neat solution to this problem by simply eradicating the animal's inside. With this view, all we know is external behaviors. We can observe sequences of movement and also, through our own behavior, manipulate the animal's behavior. This view is a modern version of Descartes' idea that animals are machines without souls. Untold harm has been done to animals because of our ability to objectify them—to make them into things we can treat as mere objects.

Any human being who has not been totally blinded by dogma knows that cats, squirrels, mice, and deer are all creatures that experience the world. This knowing is not intellectual; it is a kind of felt knowing based on the direct interactions we have with animals. The cat looks at us when we walk by and purrs when we stroke it; it raises its tail, arches it back, hisses and focuses intently on the little puppy trying to get near it. The gaze, the utterances, and the movements of the body are all gestures that are expressive of the animal itself.

To use the phrase of philosopher Thomas Nagel, "There is something it is like to *be* that organism."[1] Each animal has a perspective, a point of view through which it lives in the world. When we observe an animal, we observe how it is living out this perspective, how it is living its unique way of being.

We may never be able to take on this perspective as a first-person ("first-animal") experience. But that doesn't mean the inwardness of the animal is an impenetrable black box. It is true that we create a problem for ourselves when we imagine the inwardness of the animal as totally distinct and other from what we call the body. But what we actually confront in our experience of animals is the ensouled living body. We can't talk meaningfully about the animal's behavior, for example, if we don't include how it attentively and selectively relates to the world around it. This we can call the animal's intentionality. My cat reacts very differently to me than it does to our little puppy. That is its perspective, its way of relating, how it shapes its existence by interacting in different ways with different things.

What we can do is to carefully observe an animal's behavior and the concrete context of different kinds of behavior to gain an understanding of its specific intentionality. But we can't fully penetrate this behavior without attending to *how* it moves and the way this movement is shaped through the form and activity of its various organs. We can discover how the shape of the wings and the configuration of muscles allow a bird to fly in a particular way. The specific form of the wings and muscles are the bodily expressions of the eagle's or the chickadee's whole style of existence, which includes its intentionality and behavior.

The point is to build up vivid pictures of the animal from as many sides as possible. By continually immersing ourselves in concrete observation and then connecting our observations to vivid inner images, we enter into a conversation with the animal. The animal begins to show itself.

The Star-Nosed Mole

A mole is a highly specialized creature. Its limbs hardly seem to protrude from its barrel-shaped, compact body, and externally one doesn't see a neck—the head appears as a tapered extension of the trunk (Figure 3.1).

FIGURE 3.1. The star-nosed mole (*Condylura cristata*). The adult mole's body, without tail, is about four and one-half inches long (11-12cm.).

Noting, as he always did, the relations *between* the parts of an organism, Goethe remarked, "The neck and extremities are favored in the giraffe at the expense of the body, but the reverse is the case in the mole."[2] The cylindrical form is ideal for moving through the soil. And the dark fur—a star-nosed mole is virtually black—can bend in all directions, so that it will lie flat whether the mole is moving forward or backward through its tunnels.

When you look at the mole's forelimbs more closely, you see the proportionately enormous clawed paws with which the mole digs its subterranean tunnels. The paws are held with the large, broad palm turned outward and next to the head. When the mole digs, it stretches the paw forward, digs into the soil, and then scoops it away to the side. It is the kind of motion we make in swimming the breaststroke.

Just as the form of its body and the form and activities of its limbs express the subterranean habitat in which the mole lives and the way, as a digging mammal, it interacts with and shapes its environment, so do its senses and sense organs. Moles have very small eyes that are functional, but in many species they are not discernable until one pushes aside the fur. Moles also have no external ears (enhancing the smooth, barrel-shaped form of the body). Through field and experimental observations, it is evident that neither sight nor hearing is its primary sense, which isn't terribly surprising for an animal that lives most of its life in the earth in dark tunnels.

The star-nosed mole (*Condylura cristata*) is a particularly fascinating species of mole. Its eyes are usually visible, and its forepaws are proportionately not as large as in many other moles. The star-nosed mole spends at least some time foraging above ground. But it also spends a good deal of time in water, using its limbs as paddles. Often its tunnels lead into ponds, streams or wetlands. (This species is found in the northeastern United States and southeastern Canada.) One finds the tunnels of star-nosed moles near and around wet areas, so they are often digging through mucky soil.

But what makes the star-nosed mole stand out most is the star at the front of its snout. The star, which is less than a half an inch in diameter, consists of 22 rays surrounding the nostrils, just above the mouth (see Figure 3.2). Since no other animal has such a starlike appendage, it has intrigued researchers. Here I draw primarily on the research of Kenneth Catania at the University of Vanderbilt (see references).

When the mole is digging, the rays of the star are in constant movement. They contact and probe around, palpating (but not digging into) the earth. While digging through the soil the mole lays bare its food, which consists primarily of earthworms, but also insect larvae and other small soil invertebrates. Catania noticed that when the forepaws touch earthworms while it is burrowing, the mole does not stop, grab, and eat the worm. But when a ray of the star comes into contact with a piece of earthworm, the mole orients the star around the food and touches it rapidly numerous times with the rays. It then moves its snout so that the small and inconspicuous eleventh pair of rays, which lies just above the mouth, touches the food and engulfs it. The animal always goes through this sequence before it eats anything. As Catania describes,

> The star moves so quickly that you can't see it with your naked eye.... Scanning its environment with a rapid series of touches, a star-nosed mole can find and eat five separate prey items, such as pieces of earthworm we feed them in the laboratory, in a single second.[3]

FIGURE 3.2. The head and large clawed forepaws of a star-nosed mole emerging out of the soil. The clearly visible star consists of eleven pairs of rays ordered around the nostrils (dark spots near the center of the star). The top middle pair is very small and inconspicuous, appearing as two knobs. The eleventh pair, which is the center of focal touch sensitivity, is the small pair at the bottom middle of the star. Note also the long tactile whiskers (vibrissae) that radiate out behind the star. (Drawing from a photo in Catania 2000, p. 66)

Although it surrounds the nose, the star is not part of the mole's sense of smell, and the mole doesn't use it as an appendage to grab and handle food. Whether the star helps the mole perceive temperature differences or maybe even the electrical properties of its environment is still open to debate.[4]

The development of this mobile and sensitive appendage is correlated with other changes in the face and jaw. The muscles to move the star are highly developed, while the jaw muscles have weakened and the lower jaw (mandible) is very thin boned. This is a telling example of what the great French comparative anatomist Cuvier called the law of the correlation of parts. In Goethe's words, "Nothing can be added to one part without subtracting from another, and vice versa."[5]

Living in a Tactile World

When Catania looked at the fine anatomical structure of the rays, he discovered an extraordinary organ of touch. Under a microscope, the star's surface looks like a honeycomb of about 25,000 little domelike structures called Eimer's organs, each consisting of three different types of sensory receptors — for detecting vibrations (such as the wiggling earthworm), pressure on the skin, and the texture of objects. Catania discovered that the star

> is supplied with more than 100,000 large nerve fibers. By comparison, the touch receptors in the human hand are equipped with only about 17,000 of these fibers. Imagine having six times the sensitivity of your entire hand concentrated in a single fingertip.[6]

So we have to imagine an extremely fine sense of touch concentrated in the star. In imagining this tactile world of the mole, we must strip away what is so familiar to us—our colorful and airy world of sight and hearing. We can picture ourselves in a dark, quiet, enclosed space where the surface of our body touches myriad objects. Since our sense of touch is most refined in fingertips and tongue, we can imagine concentrating our perceptions of weight, texture, and temperature through these organs. In this way we can begin to acquaint ourselves with a tactile world that normally stands in the shadows of our more dominant and focal visual and auditory experiences. (When I was about thirteen years old, I spent a good deal of time with friends crawling around in the rain catchment pipes that ran under streets of the town I lived in. They got smaller and smaller, to where you couldn't turn around any more. We'd lie there for a while and then wriggle our way back out. I shudder at the thought of having done this and wouldn't go two feet into such a pipe today. But at that age I lived, evidently, in a different consciousness—I won't necessarily say I was more molelike—that allowed those journeys into a narrow tactile space.)

Sensory Experience: Focus and Background

In his descriptions of the mole's star, Catania makes an illuminating comparison to other senses. When we use our eyes, we are continually active in at least two ways. While reading, for example, we focus on particular words and yet at all times this center of attention is seen within the larger context of the background of sentences and words that come before and after. Or when we're walking through a landscape, we focus on a bird flying through the bushes, but we also see the surroundings as a backdrop.

This contrast between focal attention (foreground) and peripheral background has its anatomical correlate in the fovea of the retina, which is strongly innervated and used for focusing on objects, and in the periphery of the retina, which helps us see the overall surroundings.

Catania suggests that the way moles use their star reveals a similar two-fold awareness of the environment. Continuously moving its outspread rays in all directions, the mole probes its tactile world. It's a bit like us going out with our eyes wide open but unfocused; we're open to what comes. When the mole comes into contact with potential food, it focuses on it by touching it with the eleventh pair of rays. Only then is it eaten.

As in the eye, the contrast between focal sensitivity and more general awareness finds its physical reflection in the anatomy and physiology of the star and the mole's nervous system. The eleventh pair of rays has a higher concentration of nerve endings than the other rays. The nerve fibers lead to the sensory field of the cerebral cortex. Brain studies on other animals and the human being have shown that more sensitive organs or tissues are connected with larger areas of the sensory field. "Brain space" is not a function of the size of the organ, but of its sensitivity. The fovea of the retina has more brain space than the rest of the retina, just as the sensory field for the human tongue is much larger that that of the trunk of the body excluding the limbs.

Catania and his colleague, Jon Kaas, confirmed this correlation in the star-nosed mole: by far the largest part of the sensory field is represented by the star.[7] The next largest area is for the paws, followed by the area connected to whiskers behind the star on the snout. What surprised them was that the brain field for the star was actually divided into raylike projections—a star within the brain! The brain ray for the small eleventh ray was much larger than any of the others, indicating how the animal focuses its sensitivity through this part of the star organ. So even in the fine structure of the brain we find the poles of focus and background in sensory activity represented.

Although we have no comparable organ, we can gain an idea of the star as a sensory organ since we are also sensory beings. In sense activity, we have the interweaving of open exploration and centered focusing. With sight, we know this interplay well, since sight is our primary sense through which we focus our attention on the world, but without the background, without the general sensory openness that allows us to take a whole picture, we wouldn't see anything. This mutual relation between focus and background metamorphoses in each particular sense and in each animal with its specific configuration of organs and senses.

For example, the sense of smell is usually not a focal sense in human beings. But it is for a fox. Once I was following the tracks of a red fox in the snow and came upon a slightly raised spot with some twigs of a bush sticking through the snow upon which it had urinated. This is a fox's scent marking. It smelled musty and somewhat like a mild form of skunk scent. I realized how often I had smelled this scent on walks, but never put it into any context. I found a few more scent markings and had a flash of what it must be to be a fox, nose low to the ground, wandering through a world of scents and at the same time putting scents out into the world.

To enter the world of the star-nosed mole we must leave behind so much that is familiar to us. It is a dark world of continuous contact with the cool moist earth, the mole digging through that earth with large, powerful paws, and an organ out front, in continuous movement, probing and discriminating. A "mole's eye view" of the world is a view through touch.

CHAPTER 4

Where Does an Animal End?
The American Bison

On a hike in the Hayden Valley of Yellowstone National Park, my wife and I could see in the distance a few bison grazing. As we came closer, we noticed that they were bulls. They were very close to our trail. We figured it was not wise to walk between them, so we made a wide arc around them on the sagebrush-covered hillside. It didn't seem that they paid much, or any, attention to us. We continued on our way and they continued grazing.

On such an occasion I have a clear and distinct awareness of where the bison are and where they end. They are over there and I am over here. I can judge the distance between us. I don't question where the bison ends—it ends at the boundary of its massive body, and that body is over there. This knowledge of a physical boundary gives me a certain feeling of security. I can adjust my distance to the animal accordingly.

As true and as important as this may be, it is certainly not the whole story. When I am observing a group of bison cows and their calves, and one of the calves looks up and gazes at me, where does that calf end? And where do I end, the "I" that is attending to the calf trotting behind its mother? When the calf sees me, she is with me. When I perceive a bison's dark glistening eye, a young bull rolling in the dirt releasing a cloud of dust, or a bison swimming across a river with only its head above water, I am with those bison. I am here and I am there. The bison extend into me and I into them. We intermingle.

Figure 4.1. Bison bull in Yellowstone National Park.

So while the knowledge of an animal's physical boundary—and of my own physical boundary—is essential in my navigating through the world, it is a limited perspective. I have in mind only the spatial aspect of the animal and myself. But when we perceive one another and respond to one another we are in those moments not separate. We extend beyond our physical boundaries. It is no longer a simple matter to say where an animal ends. That is what I want to explore.

The Embodied Center

An individual bison is easy to recognize as such, whether a large old bull, a young calf, a ruminating cow, or a spike-horned yearling. Each has its own distinct boundary and moves as a unitary creature without fusing with its environment in such a way that it would become indistinguishable from it. That only happens when the animal dies, its body decays, and it becomes part of the soil.

Of course, this distinct physical form with all its bulk does not exist on its own. Every bison needed parents to reproduce and a mother in whose body it developed. A bison fetus developing in its mother's womb is still wholly connected with her and part of her. At birth the calf becomes a separate body and center of independent activity. But even then this separateness is only partial. It feeds on its mother's milk and, after weaning, on grass. This "separate" being will always need the sun, air, water, grass, the solid earth to move on, and more, in order to exist. Without these it would not be.

But it is also the case that these conditions for its existence do not "explain" the bison, that is, make understandable its impressive size, its unique shape, or its manifold habits. Every bison is a specific center of activity, even though this center could never exist on its own. It needs a periphery from which it draws and to which it gives, a periphery that it incorporates, transforms, and excretes in order to remain itself until it dies.

Before I venture into building up a picture of this manifold periphery that supports the center—or perhaps I could say, a picture of the "peripheral bison"—I want to consider the unique physical presence that every bison embodies as it wanders over the grasslands.[1]

The bison is the largest of all terrestrial mammals in North America. Mature bulls can weigh over 2,000 pounds and stand six feet high at the shoulders. Cows are considerably smaller, weighing "only" up to around

1,100 pounds and standing four to five feet high at the shoulders.

A striking feature of the bison's form is its massive front half that contrasts starkly with its relatively slender hindquarters (see Figure 4.2). The bison carries its head low to the ground, and long and shaggy fur covers the head. The pointed beard—which can nearly touch the ground when the bison stands and walks—emphasizes the downward orientation of the head. The shaggy fur extends over the shoulders and the upper part of the front legs, and then ends abruptly. This fur is not only longer than on the rump, but also "two to five times thicker than the hair on the slimmer hindquarters."[2] During the winter the fur is more uniform as the hair grows longer and thicker in the rear half

Figure 4.2. Mature bull; National Bison Range.

Figure 4.3. Pronghorn; Yellowstone National Park.

of the animal, but the contrast between front and rear is still apparent. Bison are quite comfortable in cold climates—"A square inch of buffalo skin has ten times as many hairs growing from it as a square inch of [domestic] cow skin."[3]

To appreciate its unique form, we can compare it with that of the pronghorn, a fellow grazing mammal of the prairies and grasslands (see Figure 4.3). The pronghorn, which is much smaller and lighter, carries its head high and its horns emerge vertically out of the skull. Its long neck holds the head above the body, and long slender legs carry its barrel-shaped trunk.

In contrast, the bison's emphasis is forward and down. It has short legs and a short thick neck, and holds its head below shoulder height. From the wide, low-held head, the neck rises into the hump above the shoulder. This hump is not, as it is in a camel, a cushion of mainly

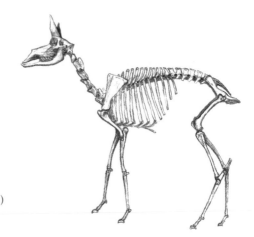

FIGURE 4.4. Skeletons
of pronghorn and bison.
(Bison: modified from
Hornaday 1894, Plate XXI)

fat tissue. It consists of the long processes of the thoracic (rib-carrying) vertebrae (see Figure 4.4). The heavy, low-held head is supported by strong muscles that are connected to these long processes. No other hooved mammal has such long processes or carries its head so low to the ground. It is not surprising that when two bulls spar, they butt with lowered heads that nearly touch the ground (see Figure 4.5).

A bison calf hardly resembles the adult—you could easily mistake it for the calf of a domestic cow (see Figures 4.6 and 4.7). It neither holds its head so low, nor does it have a hump or the marked distinction in fur between

FIGURE 4.5. Two young bulls sparring at a wallow; National Bison Range, Montana.

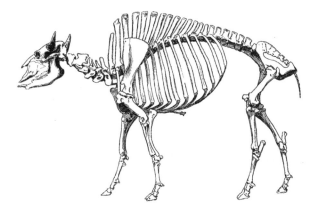

the front and back of the body. These characteristics develop over time in concert with one another as the animal grows. The calf's coat is a much lighter reddish brown and becomes darker with age.

FIGURE 4.6. Newborn calf with mother; Yellowstone National Park.

FIGURE 4.7. Older calf with mother; National Bison Range.

The bison's horns grow throughout its life (see Figure 4.8). They start as small protrusions, become within a year fairly straight spikes, and then curve upward and grow thicker over the coming years. With greater age, the horns begin to curve inward at the tips and become very thick at the base. This growth and the in-turning curve increase the impression of concentration and power in the head. All the while, the head is growing and changing its shape, becoming much broader and more massive (see Figure 4.9). This gradual growth and lowering of the head is complemented by the thickening above the shoulders through the growth of the processes of the thoracic vertebrae and neck muscles. During this time the fur on the head, neck, and shoulders becomes much longer and thicker. The unique gestalt of the adult bison takes shape.

The bison radiates concentrated force. When an adult bison walks, you witness the gravity of every step coming to earth. But for all its bulk, a bison can run fast and leap when needed. When watching a group of bison gallop down a slope, you behold an immense forward thrusting energy barreling through the landscape

Sensory Expansion and the Herd

An animal's senses mediate its perceptions of its own body and also allow it to expand beyond its physical boundary. The sense of touch is spread over the entire body and gives the animal both a perception of this boundary and an awareness of other solid objects that it comes into direct contact with. Bison love to roll and rub their bodies on bare ground, creating momentary dust clouds as well as lasting depressions in the ground called wallows (see below). They seek contact with other solid objects to rub against—trees, bushes, or rocks. Trees on the edge of grasslands are often girdled bark-free from rubbing bison (see Figure 4.10). When telegraph poles were being erected through the prairies in the nineteenth century, they were occasionally toppled by rubbing bison, and at least one settler reported that his cabin was pushed over by a group of vigorously rubbing bison![4] Here we get a glimpse into an imposing, compactly constituted animal that seeks resistance and contact with the solid features of its world.

With its head held so low to the ground as it walks along, a bison's face is in tactile contact with the plants that it feeds on. In this sea of plants it

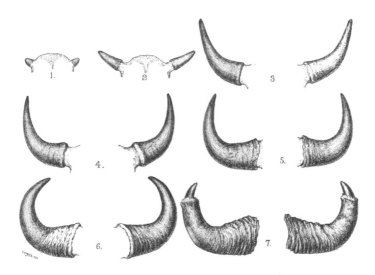

FIGURE 4.8. Horn development in the bison. **1**: calf; **2**: yearling; **3**: spike bull, 2 years old; **4**: spike bull, 3 years old; **5**: bull, 4 years old. **6**: 11 years old; **7**: old "stub-horn" bull, 20s year old. (Adapted from Hornaday 1887/2002, Plate VIII)

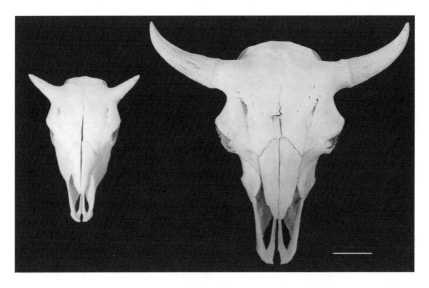

FIGURE 4.9. Skulls of a yearling bison (left) and an approximately five-year-old bull (right). Scale bar: 10 cm.

FIGURE 4.10. Trees with lower trunks rubbed bark-free by bison; Hayden Valley, Yellowstone National Park.

also orients through its keen sense of smell, discerning the qualities of the plants before it imbibes them, and then tasting them as they are briefly chewed and swallowed.

It seems that a bison can extend a mile or two outward through its senses, just as the bellowing of a bull in the rutting season can be heard from such distances. Bison detect scents that waft their way from great distances, a fact well known to scientists and hunters who want to get close to a herd without disturbing it and have learned through experience to approach a herd from the downwind side.[5]

It is easy for us to think of a herd of bison as a collection of individuals. But if you take the perspective of any individual bison, then its existence is clearly bound up with that of the herd. Through their senses of touch, smell, hearing, and vision, the members of a herd weave into one another. A cow identifies her calves in the first days mainly by smell; then she recognizes it visually, and finally by its calls. Grunts and bellows resound and carry manifold meanings for members of the herd. A bull smells the urine and rear end of females, which can tell him whether she is in heat or not. The tail is a highly expressive organ, and through vision bison can participate in the ebbs and flows of dispositions and moods that show themselves through the tail, as well as through movements of the body and head.

The herd is not an add-on to individual bison life; you can't understand the life of any individual without considering it as a herd member. Being part of the herd does not mean having a specific role or function. Rather, the animals live in a landscape of shifting relations that at times intensify and at times loosen. An old bull, for example, may spend a good part of the year by himself, but in breeding season (July and August) he will usually return to a herd and interact with other bulls—often vehemently—and mate, or try to mate, with cows.

Watching the northern Yellowstone Park herd along the Lamar River in June, you can form one picture of this extended organism of the herd. Over a thousand animals spread out and move around through this valley. Each day you view a different scene and different groupings. At this time of year virtually every cow has given birth to a calf that stays close to its mother throughout the day. It lies near its mother, and when she feeds and moves along the calf soon follows (Figure 4.11). Perhaps 20 to 40 cows with calves and yearling females and males often form a cluster that grazes, ruminates and moves together. But such groupings are not stable or fixed. Some of the cows and calves may swim across the swift current of the Lamar River and begin mingling with other cows and calves, while the remaining animals stay behind and become part of some other fluctuating cluster.

During early summer you rarely see bulls among the cows. The bulls form separate bull groups, or wander about singly, especially if they are

FIGURE 4.11. Part of a grouping of cows and calves in June in the Lamar Valley of Yellowstone National Park.

older bulls. You often see three to seven bulls—younger and older—grazing together and moving across the landscape together. Often the bulls spread out from each other and then move closer together. When they are close, younger bulls will often head butt and wrangle, only to spread out again and give themselves over to grazing.

One time I saw a lone older bull approach a bull group. When he was among the others, he moved very slowly and raised his tail into an arc. A bull from the group started grumbling. It was a tense atmosphere—as the saying goes, the air was so thick you could almost cut it. The older bull stomped with his forefeet, dug repeatedly at the dirt, and lowered his head, swishing it forcefully back and forth against a sagebrush bush. Soon the bull group moved along and the bull was alone again. From the other direction three bulls trotted down the hillside and walked toward the lone bull. In this meeting there was no palpable tension—no grumbling, snorting, stomping, or head shaking. The lone bull turned around and joined the three bulls as they moved east in the direction of the other bull group. For a while at least, the lone bull was part of a group.

These few vignettes show that the herd is neither an agglomeration of individual animals nor a group with fixed roles and functions. It is a continually shifting relational dynamic. At times, dense and focused soul spaces are created, if I may put it this way—when a calf suckles; when a cow crosses the river to meet her calf that had been running back and forth along the bank to find her; when a bull enters a group of other bulls. And at other times, the tension and attention among herd members loosens as they give themselves over more to grazing or ruminating—turning toward the plant world that sustains their lives. The relational life of the animals contracts and expands during the day and year.

The Intertwined Existence of Grassland, Bison, and Microorganisms

In 1800, millions and probably tens of millions of bison lived on the midcontinental grasslands.[6] Their range extended from Mexico in the south up into Canada, and from the Rocky Mountains into Indiana. Bison were virtually exterminated by 1890, when only around 1,000 animals were still alive in all of North America.[7]

Still, into the 1870s bison formed immense herds throughout their range. Colonel R. I. Dodge writes, for example, about traveling a distance of 35 miles in Kansas in May, 1871, when "at least twenty-five miles of this distance was through one immense herd.... The whole country appeared one mass of buffalo, moving slowly to the northward; and it was only when actually among them that it could be ascertained that the apparently solid mass was an agglomeration of innumerable small herds, of from fifty to two hundred animals, separated from the surrounding herds by greater or less space, but still separated."[8]

Bison herds moved in relation to the seasons and to food availability, wandering hundreds of miles in any given year. When bison move through the prairie they are moving through a sea of their food. A bison doesn't tend to snip off the tips of plants, but moves with its snout near ground level, tears off shoots close to the soil, chews briefly, salivates copiously, and swallows.

The grass enters a small chamber of the stomach called the rumen. This muscular chamber massages the grass by rhythmical muscle contractions, and it grows and develops through this interaction. Over time it becomes the largest chamber of the stomach, thanks to its continuous engagement with grass. In the rumen the grass is churned around in swallowed saliva and balls ("cud") of food are formed that are then regurgitated and chewed thoroughly before swallowing—a process called rumination. We ruminate on thoughts and feelings; bison ruminate on the prairie they have internalized, which eventually becomes part of their bodies.

It is no simple matter to live from grass.[9] Grass plants are remarkably tough structurally—it is no small feat of nature to create exceedingly thin, upright, stable yet resilient stems and leaves. The fibrous cell walls even incorporate silica in the form of opal as a structuring element.[10] In a way it is paradoxical that bison and many other large grazing mammals, live from grass, which is both hard to digest and, from the perspective of carnivores and herbivores, poor in nutrients.

A large animal like a bison must take in very large amounts of grass to live. As a bison feeds on grass, it is not only ingesting grasses and some other types of plants. It is also taking in microbes (bacteria, protozoans, and fungi) that live on plants, in the upper layers of soil, and on other animals. So when a calf licks its mother, or its mother licks the calf's lips, the calf is also ingesting microbes. Conversely, when a bison is feeding on grasses, it is leaving behind microbes in its saliva on the grasses. So there is an abundant sharing of microbes between animals, plants, and the soil.

Without this sharing, a bison could never digest grass. In the dark, warm, and fluid environment of the rumen, microbes find ideal living conditions. Their food is grass (and each other). Many microbes can form enzymes that break down cellulose into other carbohydrates (starch and sugars) they can utilize for their own growth. In turn, the microbes release fatty acids into the rumen that the bison can use for its own growth. Some of the microbes move into the other chambers of the stomach and in the final fourth chamber (the abomasum) they are themselves digested and provide a vital source of protein for the animal.

As the bison grows and feeds, it develops what we could call a microbial organ within the organ of the rumen, which itself develops in interaction with the grasses and microbes.[11] So the bison gathers microbes from the surroundings, including other bison, and the microbes rapidly become an integral part of the animal without which it could not live. The internal microbial ecosystem develops into an organ of the bison.

The Mutual Dependency and Enhancement of Bison and Prairie

While a prairie consists mainly of a variety of grass species, there is also a great variety of wildflowers (forbs) that are, however, not so great in number. Because bison feed preferentially on grasses, they leave many forbs ungrazed. Areas that have been grazed by bison have greater plant diversity and the plant community is more varied than in ungrazed areas.[12] (All this assumes, of course, that the bison are free to roam and are not forced to overgraze due to confinement.)

When a bison has grazed on a grass plant, the plant responds by increasing its growth. Young shoots are more nutritious than older ones, and researchers have observed that bison tend to return after a time to already grazed areas to feed on the fresh young grass. In such patches more forbs grow than elsewhere. Eventually the grass growth diminishes and the bison feed elsewhere.

When there are more wildflowers, more insects and other invertebrates thrive, and they are also present in greater diversity.[13] They are, for example, important pollinators and are connected in myriad ways with plant and other animal life. As naturalist John Muir famously remarked, "When we try to pick out anything by itself, we find it hitched to everything else in the universe."[14]

When the bison ingests and digests grasses and microorganisms, it builds up and maintains itself. In this process it produces substances that it gives back to the prairie—its urine and dung. These stimulate plant growth. Since they are provided in a localized and concentrated form, they contribute to the diversity and patchwork nature of plant distribution in the prairie. For example, plants growing in a urine patch have higher nitrogen content and are sought out by bison, which graze the patch intensely. Such intensive grazing of a small area inhibits grass growth and at the same time allows forbs to thrive.

White settlers traveling through the prairies in the 1800s noted that the trails of bison herds stood out from the surrounding prairie through their bright greenness.[15] The defecating and urinating bison left a trail of nutrients, and the manure was continually being ground up and mixed with the soil through the bison's hooves. In the spring, the plants found ideal conditions for lush green growth.

One feature of "giving back" to the prairie we might not think about is animal carcasses. When a bison dies, its body decays and, if not scattered and eaten by scavengers, becomes over time part of the soil. This soil is nutrient-rich and harbors vibrant plant life.[16]

Bison bring variety into the prairie in yet another way. They wallow. They scrape away plants with their hooves and create an area (around three to five meters in diameter) of bare soil in which they roll around.[17] Gradually the wallow becomes an oval, bowl-like depression that can be a foot deep at the center. A prairie that is home to a herd of bison is pocketed with many such wallows. The same wallow may be used by many different animals for long periods of time, but bison also make new ones. Early settlers coming onto the prairie remarked on the countless wallows—some of which can still be observed as depressions with a unique composition of plant species in prairie preserves that have had no bison activity for 125 years.[18]

A wallow contains compacted soil and holds rainwater better than the surrounding prairie. Aquatic plants can populate such oases, and wallows

FIGURE 4.12
A bison wallowing;
Lamar Valley,
Yellowstone
National Park.

FIGURE 4.13.
Bison at a wallow;
National Bison Range,
Montana.

can even serve as breeding pools for frogs.[19] By contrast, during longer dry periods only drought-resistant species germinate and develop in the bare, compacted soil of a wallow. All in all, wallows make unique microenvironments that contribute to the overall patchiness and diversity of prairie life.

When bison shed their fur, it can find its way into the nests of birds and small mammals.[20] And as bison make their way through the prairie, seeds get caught in their fur. They become walking seed bearers. One study

FIGURE 4.14. Bison and many wallows; Lamar Valley, Yellowstone National Park.

found 76 different plant species in 111 hair samples from different bison.[21] At some point the seeds may become dislodged, fall to the ground, and germinate. Bison also ingest seeds along with leaves and shoots, and often these pass through the digestive tract without being broken down. In both these ways the bison spread seeds, one more contribution it makes to the prairie that sustains it.

All these detailed observations help us to appreciate how the bison and its environment support each other, affect each other, intertwine with each other. The ecological bison spreads out way beyond its body's boundary. And while neither the bison, nor grass plants, nor microorganisms are composite creatures in the sense of being put together out of external raw materials, each of these centers of activity is unthinkable by itself and necessarily intermingles with and in part merges with the others.

Fire, Plains Indians, and the Prairie

While the large numbers of roaming bison played a significant role in maintaining and diversifying the prairie, they were not alone in doing this. One other major player was fire. As we know today, fires tend to kill shrubs and trees, while grasses are more resistant to fire since they have growing points beneath the surface of the soil and can resprout after a fire. Most of above-ground parts of grass plants that die in the fall, decompose during the following year. But a significant amount— perhaps 20 percent—remains if the grasses are not grazed or burned by fire.[22] When undecomposed litter accumulates, the grasses become less productive. So both fire and grazing keep the prairie thriving.[23]

Fires arise from lightning and from human action. Today, ecologists set fires in prairie preserves, in some cases annually and in others in three-to-five-year intervals. Historically, Plains Indians regularly set fires in the vast grasslands.[24] Human-set fires have been far more important in creating and maintaining grasslands than lightning.[25] Missionary Timothy Flint, who was traveling in 1826 in the area around St. Louis, wrote:

> I have often witnessed in this country a most impressive view, which I do not remember to have seen noticed by any travelers who have preceded me. It is the burning of the prairies. It is visible at times in all parts of Missouri, but nowhere with more effect than in St. Louis. The tall and thick grass that grows in the prairies that abound through all the country, is fired; most frequently at that season of the year, called Indian summer. The moon rises with a broad disk, and of bloody hue, upon the smoky atmosphere. Thousands of acres of grass are burning in all directions. In the wide prairies the advancing front of flame often has an extent of miles. Many travelers, arrested by these burnings, have perished. The crimson-coloured flames, seen through the dim atmosphere, in the distance seem to rise from the earth to the sky.[26]

One important effect of burning in the fall ("Indian summer") was that the following spring grass grew vibrantly in the burned areas and attracted bison to these "grazing lawns," as they have come to be named. Native tribes could expect to find bison herds in these areas, which often became their spring hunting grounds.

Native tribes also used fire as an aid in hunting bison. Here is a description by the French Jesuit Pierre de Charlevoix from the early 1720s:

In the Southern and Western Parts of New France, on both
Sides the Mississippi, the most famous Hunt is that of the
Buffaloe, which is performed in this Manner: The Hunters range
themselves on four Lines, which form a great Square, and begin
by setting Fire to the Grass and Herbs, which are dry and very
high: Then as the Fire gets forwards, they advance, closing their
lines: The Buffaloes, which are extremely afraid of Fire, keep
flying from it, and at last find themselves so crowded together
that they are generally every one killed. They say that a Party
seldom returns from hunting without killing Fifteen Hundred
or Two Thousand. But lest the different Companies should
hinder each other they all agree before they set out about the
Place where they intend to hunt.[27]

The prairie was home of the bison and the bison, along with fire,
encouraged prairie growth. The main source of fire were the Native
peoples, who in turn lived from the bison that lived from the prairie that
thrived due to the fires that the tribes set. A truly interwoven fabric of
existence.

Serving the Needs of Plains Indians

The life of the different Plains tribes revolved around bison. Tom
McHugh, in his book *Time of the Buffalo*, describes in detail the variety of
ways in which the bison served the life of Plains tribes.[28] Virtually every
part of the bison carcass could be put to use for some purpose:

- Meat, internal organs, blood, and bone marrow were used for food.

- Whole hides, painstakingly dressed, provided robes, rugs, and walls
  for sweathouses and tepees, which required anywhere from seven to
  twenty hides. Clothes, moccasins, hats, belts, and mittens were made
  from skins.

- Rawhide, which consists of skins that have only had the flesh and hair
  removed, served as "knife sheaths, cups, dippers, kettles, mortars,
  rattles, drumheads, cradles, cages, fencing, boats, cases, shields, bridles,
  lariats, and other pieces of saddlery—including small bags that were
  wrapped around a horse's hoofs and served as shoes."[29]

- Horns were fashioned into spoons and ladles, as well as "arrow points, drinking cups, powder flasks, trimmings for war bonnets, spinning tops … heads for war clubs, cupping horns for bleeding patients to relieve infections, and—after simmering with spruce needles—a medication for sore eyes."[30] The Crow and Cheyenne made high quality bows by cutting horns into strips, piecing them together, and binding them with sinews that had been soaked in glue.

- Hooves along with the muzzle, eyes, penis, and other parts of the animal were boiled together to make glue.

- Bones became "war clubs, pipes, knives, knife handles, arrowheads, arrow-making tools, and runners for small, dog-drawn sleds." The frontal bones of the skull were shaped into hoes and spades, while the fleshing tool used in tanning hides came from the leg bones. The porous inside of the tops of limb bones could be shaped into a sanding tool and also became paint "brushes"—the pores soaked up the colored fluid, which could then be spread onto leather.

- Teeth were mainly used for jewelry and ornaments.

- Tendons (sinew) are very tough, and strands could be stripped from them to make durable threads. When twisted together, such threads became ropes, bowstrings, and bindings used for a variety of purposes such as fastening points to arrows.

- Hair (separated from the hide) became a lining for moccasins, or was shaped into dolls and balls for children. The hair strands were also braided or twisted into cords that were used, depending on thickness, as loop earrings, bracelets, halters, belts, and ropes.

- Rumen, bladder, the membranous sac around the heart (pericardium), the large intestine, and even the whole skin of a calf could be used to hold water and food.

- Gallstones provided a yellow pigment for paint, and the Cheyenne made "a black pigment by stirring cottonwood buds or ashes of burned grass into fresh buffalo blood."[31]

- Dung ("buffalo chips"), when dried, could be burned like peat. Dried dung could also be pulverized into soft fibers and pressed into pads that were used as absorbent baby diapers.

How integral the bison—after death—was to the life of the Native people! The bison sustained them as food, and its recrafted manifold parts enveloped them and made their day-to-day life possible. The transformed bison was an active and essential component of the Plains tribes. Clearly, in this sense, a bison does not end when it dies. Its life as an individual organism is gone, but through its body parts the bison integrates into a whole new life world—that of the Native people.

The Bison as Spirit Being in the Culture of the Plains Tribes

What I've ignored in the above description is the role bison played in what we would today call the spiritual life of the Plains Indians. To separate day-to-day life from the spiritual is something we do in modern Western cultures. This separation was not present in the life of the Plains Indians.[32]

A bison hunt, for example, was not just a matter of killing an animal for its useful products. It was an extension and expression of the Plains Indians' relation to bison as physical-spiritual beings. Most of us today who grow up in Western cultures and use an animal or plant for food or other needs do not feel a connection to some larger spirit nature of the animals or plants we consume. But this was very different for the Plains Indians, as this story from the Blackfoot (Nitsitapii) tribe reveals:

> Long ago, in the winter time, the buffalo suddenly disappeared. The snow was so deep that the people could not move in search of them, for in those days they had no horses. So the hunters killed deer, elk, and other small game along the river bottoms, and when these were all killed off or driven away, the people began to starve.
>
> One day, a young married man killed a jack-rabbit. He was so hungry that he ran home as fast as he could, and told one of his wives to hurry and get some water to cook it. While the young woman was going along the path to the river, she heard a beautiful song. It sounded close by, but she looked all around and could see no one.
>
> The song seemed to come from a cotton-wood tree near the path. Looking closely at this tree she saw a queer rock jammed in a fork, where the tree was split, and with it a few hairs from a buffalo, which

had rubbed there. The woman was frightened and dared not pass the tree. Pretty soon the singing stopped, and the I-nis'-kim [buffalo rock] spoke to the woman and said: "Take me to your lodge, and when it is dark, call in the people and teach them the song you have just heard. Pray, too, that you may not starve, and that the buffalo may come back. Do this, and when day comes, your hearts will be glad."

The woman went on and got some water, and when she came back, took the rock and gave it to her husband, telling him about the song and what the rock had said. As soon as it was dark, the man called the chiefs and old men to his lodge, and his wife taught them this song. They prayed, too, as the rock had said should be done. Before long, they heard a noise far off. It was the trample of a great herd of buffalo coming. Then they knew that the rock was very powerful, and, ever since then, the people have taken care of it and prayed to it.[33]

This story of the first *inisikim*, or buffalo rock, has been passed down, in variations, until the present. When an Indian found an *iniskim*, it became

strong medicine, and, as indicated in some of these stories, gives its possessor great power with buffalo. The stone is found on the prairie, and the person who succeeds in obtaining one is regarded as very fortunate. Sometimes a man, who is riding along on the prairie, will hear a peculiar faint chirp, such as a little bird might utter. The sound he knows is made by a buffalo rock. He stops and searches on the ground for the rock, and if he cannot find it, marks the place and very likely returns next day, either alone or with others from the camp, to look for it again. If it is found, there is great rejoicing.[34]

The *iniskim* became part of a sacred bundle that was important in ceremonies to call the bison. So we see here a powerful story culture in which an object has a latent force that can be freed when people give attention to it and use it in an appropriate way. There is, in this way of being in the world, no such thing as a merely inanimate object.[35]

The sacred pipe of the Lakota Sioux was brought to the tribe by White Buffalo Cow Woman, and she instructed them about its significance and use. When she was leaving them, she turned first into a "young red and brown buffalo calf," then into a "white buffalo," and finally into a "black buffalo. This buffalo walked farther away from the people, stopped, and after bowing to each of the four quarters of the universe, disappeared over the hill."[36] The pipe itself was a kind of microcosm of the universe. White Buffalo Cow Woman told the tribe:

> With this sacred pipe you walk upon the Earth; for the Earth
> is your Grandmother and Mother, and she is sacred. Every step
> that is taken upon Her should be as a prayer. The bowl of this
> pipe is of red stone; it is the Earth. Carved in the stone and facing
> the center is this buffalo calf who represents all the four-leggeds
> who live upon your Mother. The stem of the pipe is of wood, and
> this represents all that grows upon the Earth. And these twelve
> feathers which hang here where the stem fits into the bowl are
> from *Wanbli Galeshka*, the Spotted Eagle, and they represent the
> eagle and all the wingeds of the air. All these peoples and all the
> things of the universe are joined to you who smoke the pipe—all
> send their voices to *Wakan-Tanka*, the Great Spirit. When you
> pray with this pipe you pray for and with everything.[37]

The pipe is very "*wakan*," meaning powerful, holy, and sacred. It is not merely an outer symbol of the universe. It embodies the universe, is a presence of the universe. When we hear, for example, that the buffalo calf carved in the pipe "represents" all four-leggeds, we might think of it as an illustration or design element. But for the Plains Indians each feature was a re-presencing of a universal power. The pipe was used in at least seven different ceremonies.[38] These ceremonies were human enactments—often involving purification—to create renewed relations in the tribe and in the rest of the universe to the power and sacredness of the world, to *Wakan Tanka*.

In this sacred world, the bison and the human being are people. As part of the instructions for preparing the sun dance ceremony, the tribe is told:

> You should cut from rawhide the form of *tatanka*, the buffalo.
> He represents the people and the universe and should always be
> treated with respect, for was he not here before the two-legged
> peoples, is he not generous in that he gives us our homes and our

food? The buffalo is wise in many things, and, thus, we should learn from him and should always be as a relative to him.[39]

Or in the Blackfoot story of Scarface, the sun asks: "Which one of all the animals is most *Nat-o'-ye* [having sun power, sacred]? The buffalo is. Of all animals, I like him best. He is for the people. He is your food and your shelter."[40]

And the bison could also be a teacher. The young Iowa boy, Lone-walker, followed his father and other men on a buffalo hunt and was weeping, since he too wanted to hunt.

> In the distance he saw them shoot a buffalo bull, a small one, and leave it lying there while they passed on. Just as he was passing the carcass, sobbing and crying, the bull spoke to him. "Oh, so it's you, Lone-walker? I'm glad you came, for I've recovered and am just about to get up again. Now I'm going to tell you what to do from this time on. You must skin me over the forehead, taking my horns and a strip of fur down over my backbone to my tail, and you must use me in doctoring. Also take a piece of flesh from my leg, dry it, and pulverize it. Take some of my back fat to grease yourself and the wounds of your patients. Next remove my dew claws and make them into a rattle. You have been trying to dream something, so today I'll show you what we buffaloes will give you and you may hereafter do to your own people as we do to ourselves. This doctoring will be called *tce!hówe*, the buffalo's ways."
>
> Then the buffalo taught him the roots and herbs they used to heal the sick. They were especially potent for broken bones and wounds. He showed the lad how to use splints in binding them up and he taught him the potent buffalo songs, and what preachments and prayers to make.[41]

Extending beyond its remarkable and crucial material service to the Plains tribes, the bison inhabited their souls and became part of their actions. The bison was a fellow spiritual person, a guide, and a teacher who taught from the inside out and served materially from the outside in. The bison extended into the very core of a human culture. Or we could also say, a human culture stretched out into the many-layered being of the bison.

Can the Bison Expand Again?

The rich coexistence of Plains tribes and bison was brought to an end during the course of the nineteenth century. The Plains tribes were being decimated by illnesses (such as smallpox) carried by Euro-Americans and by the military campaigns against them. Concurrently, the population of bison dropped from millions to a few hundred by the beginning of the twentieth century. Drought, increased hunting by Native tribes on horses, exotic diseases, and especially the market for bison fur were reducing bison numbers into the latter half of the nineteenth century. The wanton slaughter of bison by hunters and soldiers came in full force in the decades after the Civil War, when railroads extended into the western plains. The power of greed ("the market") and the desire to open up the continent for Euro-Americans by ridding it of Native Americans and their source of life—the prairie and the bison—drove the decimation.[42]

The Plains Indians foresaw in dreams and visions this tragic interweaving of their destiny with that of bison. For example, Black Elk, the Sioux holy man, states:

> A long time ago (about seventy years) there was an Indian medicine man, Drinks Water, a Lakota, who foretold in a vision that the four-leggeds were going back into the earth. And he said in the future all over the universe there shall be a spider's web woven all around the Sioux and then when it shall happen you shall live in gray houses (meaning these dirt-roof houses in which we are living now), but that will not be the way of your life and religion and so when this happens, alongside of those gray houses you shall starve to death.[43]

The Plains Indians' deep connection to the bison came to vivid expression when Plenty Coups, a Crow tribal leader, said as an eighty-year-old in the late 1920s, "When the buffalo went away, the hearts of my people fell to the ground, and they could not lift them up again. After this nothing happened."[44]

After the near extermination of the bison by 1900, bison populations gradually grew again. Today there are around 20,000 bison in 62 different conservation herds in North America and around 400,000 in herds raised for commercial use (mainly meat).[45] Even the few free-roaming herds, such as the one in Yellowstone National Park, are not truly free to roam wherever they go, since outside park boundaries they can be hunted.

A question weighs heavily on anyone who gains a sense of the full life of a bison—its centered and its peripheral nature as I have tried to portray it here. Can we human beings choose to speak for the bison as they have become part of us, and provide conditions for them to expand again, on earth, into a larger world of relations?

We cannot simply reverse the tragic contraction of the bison. But we can move forward to a human culture that once again acknowledges the value of the expansive creatures that animals are. And we can work—despite all obstacles—to provide large areas where they can become agents in the revitalization of landscapes.[46] A variety of initiatives—in Native tribes, NGOs, and government—are working in this direction.[47]

One initiative is that of the Intertribal Buffalo Council, which was founded "for the purpose of restoring buffalo to Tribal lands to promote and rekindle the spiritual and cultural relationship between Tribal people and buffalo, to promote ecological restoration and to utilize buffalo for economic development."[48]

Fred Dubray, one of the Council's founding members, had a conversation with historian Dan Flores about the Council's idea of bringing the buffalo back to reservations. An elderly Lakota woman approached them and said, in effect: "Best you ask the buffalo if they *want* to come back." They performed a ceremony and asked the buffalo. They want to come back, the ceremony revealed, but they don't want to come back and be cows. They said that they want to be buffalos; they want to be wild again.[49]

A Many-Layered Being

How expansive animals can be today is largely in the hands of human beings. A crucial step is to expand our awareness to recognize and understand animals in both their centered and peripheral aspects. In this endeavor, we can discover different qualities and kinds of centers and peripheries. Physical centeredness is not the same as the centeredness of life processes, just as the web of physical relations is not the same as the web of living relations.

As a *physical entity*, a bison's robust body has a fairly clear boundary that sets it apart from other bison, and from the ground, rocks, trees, streams, and other features of its surroundings. We can view it as a thing

Figure 4.15. Bison in Yellowstone National Park

among things, each filling its own space and having a boundary. This is what allows us to say, "Look, that bison over there is so much bigger than the one up on the hillside." We can count the members of a herd. When we focus on this centeredness, we consider a bison as an independent entity, ending at its physical boundary.

But every physical thing is also bound up in relations with the larger physical world. As a spatial entity, a body needs space around it. A space-filling, material body rests on the ground that supports it. Bodies don't exist in a vacuum. As a physical being, a bison stands on the earth, and is subject to gravity, to changing temperatures, and much more. It is a member of the physical earth.

As a *living being*, a bison is also centered. Just think of its ability to generate a constant body temperature, which remains around 98°F (36°C) despite exposure to continually fluctuating outer temperatures. The constancy of body temperature is not a "thing"; it is an ongoing achievement of the life of the animal. As a living being the bison is activity. All its parts and characteristics are being actively built up, maintained, and, if needed, broken down. In the course of its life, the bison develops new characteristics, grows, and, perhaps, reproduces. All this is part of its centered life. So while we can say of a physical entity, such as a rock, that it remains over time, of an organism we must say that its identity over time is "not the inert one of a permanent substratum, but the self-created one of continuous performance."[50]

But this life could not exist—the bison could not be a living agency—without a supporting periphery. The bison feeds on the grasses and needs to digest plant life to continually maintain its own life. An organ forms in the rumen out of the microbes that the bison has taken in from its environment. It exchanges gases with the atmosphere. Its life exists as the intertwining of active center and abundant, receptive, energizing periphery. When we think of the living bison, its "boundary" is not a physical partition; it is an active, selectively permeable, and dynamic interface. And as we have seen, the life of the bison, for example in feeding, spreads out into the life of the environment that supports it. The living prairie is, in a sense, the peripheral half of the bison's life.

We experience the bison as a *sentient being* when we attend to its behavior. We observe how it notices when we come too close, how it walks toward a new grazing patch and begins feeding. We see the difference between its gaze when ruminating or when sparring with another animal. We realize that the various grunts and bellows have meanings in the relations among the members of the herd. All such activity radiates out from the bison as a centered, attentive being. Sentience is an expanding and contracting agency by which the animal opens through its senses to participate in a specific world of qualities. The bison expands into a wide world through senses such as hearing, sight, and smell and draws more into itself when tasting, ruminating, and digesting. It is with what it senses.

Every movement the animal makes embodies sentience, whether flicking of flies with its tail or swimming across a river. Every part of its body is ensouled, but the sentience itself—the soul of the animal—is not, just as the animal's life is not, a "thing" that can be localized. We need to become comfortable with the seeming paradox that sentience is both everywhere and nowhere. It cannot be grasped by a mentality that only wants to deal with what can be clearly circumscribed as a thing or a causal chain of discrete events.

When interacting with its fellow herd members or when fleeing a predator, the bison is living in its "soulscape," a world of relations that belong to it. A bison's sentient periphery is not the same as an eagle's or a mole's, though they may intersect. So while the physical bodies of different kinds of animals are clearly distinct, their sentient lives can intermingle and do so every time two animals interact. So through its sentience the bison spreads as far as its senses and limbs will take it.

Finally, the Plains tribes clearly experienced the bison as a *spirit being*. For them it was a great being that is embodied in every single member of the species. This being can interact with other spirit beings—human beings, other animals, plants, or rocks. This is the most expansive—and most elusive—aspect of the bison. Ceremonies and the attainment of special states of consciousness were necessary for the Native peoples to experience the bison as spirit being.

As a child of Western culture, I don't have an experience of the bison as a spirit being in the way the Plains tribes describe. And yet, I sense that they have encountered something real. I have a strong urge to reconnect with something larger and deeper that I intimate but don't yet fully experience. So my striving is to find ways to reconnect that feel authentic and respectful.

As I mentioned earlier in the conclusion of Chapter 2, a guiding question for me in my studies of animals is, "Who are you?" I engage with the many details of an animal's being through the study of morphology, physiology, ecology, behavior, and evolution. But I don't want to get lost in myriad facts or in a particular perspective. I don't want to "explain" an animal or its features. In getting to know the animal through its manifold features, I let it come alive in my imagination, and then I get glimpses of its unifying character. When the unifying character of the animal begins to show itself, I can speak of the *nature* of an animal. This nature is what reveals itself through all the other layers of physicality, life, and sentience as the specific way of being of an animal species. It is the bison in the bison, the sloth in the sloth. I could call this the spiritual nature of the animal, and thereby give expression to my sense that I am approaching, from a very different starting point, what the Native people called the spirit being of the animal.

If we take the fact seriously that an animal is both a centered and a peripheral being—that it does not end at its physical boundary—then we can see that a landscape, a soulscape, and a spiritscape belong to the animal. This peripheral being of the animal is not so tangible as the robust, centered animal we encounter as a bodily presence. And we can all too easily overlook the larger being of animals since they are adaptable and resilient and can therefore live under conditions in which peripheral relations have been reduced and made poorer. Animals can live in confinement as long as certain basic needs are met. But an animal is not a whole animal when it is not embedded in the rich world of relations with other beings and the earth.

CHAPTER 5

The Intertwined Worlds
of Zebra and Lion

In the Savannah

TOWARD THE END OF A LONG DAY observing animals, we were perched in our Land Rover in Moremi Game Reserve in Botswana. We were at a border between the dry bush savannah and the moist inland Okarango Delta with its waterways and tall grasses. The air was cooling and the landscape was bathed in the golden light of the setting sun. In our looking and listening, we expanded into this landscape and became happily lost in what we witnessed.

Swamp-loving antelopes, called lechwes (*Kobus leche*), were grazing near the high grass area. A lone hippo walked into a small pond and submerged itself. Other unseen hippos provided a bellowing chorus. Behind us a herd of elephants bathed in the mud; a young one rolled around, while others flung mud onto their backs with their trunks. Not far from them, zebras grazed, moving slowly along with heads lowered to the short grasses.

As I looked ahead, I saw a female lion emerge from tall grasses. She headed across the short brown grassland, advancing slowly but in a fairly straight line. The lechwes nearby continued to graze, apparently undisturbed by their main predator walking by. Soon a second lioness appeared and went in the same direction. She seemed to have as little interest in the potential prey in her vicinity as they had in her.

With binoculars I saw a third lioness lying on an old, rounded-off termite mound surrounded by tall grasses. She peered intently in the direction in which the other two lionesses were walking. After a time, she looked in other directions, descended from her vantage point, and moved around in the tall grass. If we hadn't already spied her on the mound, we would never have discovered her, so well did she blend in with the yellow-beige grasses. She moved to the edge of the tall grass and crouched down low. A male lechwe was walking toward her as he grazed. When he was about 40 feet away, she darted out and made a brief sprint as the lechwe turned and fled in the opposite direction. The lioness soon stopped running. From our human perspective, we thought: not a terribly energetic attempt at a kill. But maybe it wasn't an attempt. The lioness moved off in the direction of the other two lionesses, which had disappeared into another area of tall grasses.

As all this was happening, the hippos were bellowing, and the sky filled with starlings that were gathering and flying toward the high grasses bordering the waterways. Their loud and high-pitched calls filled the air.

The intensity of colors and sounds, and the variety of forms, movements, and activities made a lasting impression. Reflecting back on this experience, I was struck by the "different worlds" present in one place—hippos, starlings, zebras, elephants, lechwes, and lions. Each is so different from the others, and each is in its own right a commanding presence. The different animal species seemed to give little heed to the others around them. Only when the lioness darted toward the lechwe did momentary interaction disrupt the seeming independence of species. What we call savannah, with its varied and changing composition of grasslands and trees, is home to all these animals, and yet each species pursues its own existence in keeping with its very specific way of being. What can we say about these different ways of being and their interactions? Let's ask the zebra and the lion.

Zebra and Lion in Their Worlds

With their bold and beautifully rhythmical black-and-white-striped coat, zebras stand out as a striking appearance in the green (rainy season) or straw-colored (dry season) savannah grasslands where they live. I will be referring mainly to the plains zebra (*Equus quagga*, formerly designated as *Equus burchelli*[1]). Its range extends from East Africa down to South Africa.

As grazers, zebras live in the midst of their food—grasses of the savannah. When there is ample growing grass, zebras spend about 14 hours of a 24-hour day grazing. During the dry season, they may spend 19 hours grazing.[2] As we have seen, grass is very tough and not easy to digest. But it is abundant in the savannah, and zebras take in large

FIGURE 5.1. Adult female and young foal of plains zebra (*Equus quagga*) in Etosha National Park, Namibia. (Drawing after a photo by Angoria, Wikimedia Commons)

amounts of grass each day. Zebras prefer the stems and leaf sheaths of short grasses, but will also feed on leaves and grass seeds.[3]

With head lowered to the ground, a zebra stands and walks slowly along as it grazes. Its agile lips, its nostrils, and its large jaws are in constant movement as it clips off the grass with its large incisors. With its tongue it brings the food between its massive cheek teeth (premolars and molars), and in a rhythmical circling motion the grass is sheared, ground, and moistened with saliva before it is swallowed. When we imagine this activity occurring between 14 and 19 hours every day, we realize not only the focal nature of grazing in their lives, but also the persistent repetitive activity that flows at each moment into their interaction with grass.

We saw that the bison is able to digest grass only because of the microbes that live in its rumen. No mammal is able by itself to digest cellulose, which is a major component of grasses. All herbivores have developed some organ that provides an environment in which cellulose-digesting microorganisms can thrive. Unlike in ruminants, such as the bison or the cow, the zebra does not have a four-chambered stomach with an expansive rumen housing microorganisms. Nor does it regurgitate its food and chew cud. The zebra's digestive organ that is comparable to the rumen is an enlarged portion of the large intestine (the caecum), situated in the rear part of the digestive tract. Food passes more rapidly through a zebra than it does through a bison or comparably sized antelope and is less fully digested when excreted.[4] In compensation for its more rapid and less intense digestive process, the zebra takes in more food during the day than does a ruminant.

Zebras spend most of the day standing and walking—approximately 20 hours a day. They can sleep while standing. Their grazing lives demand endurance in head, digestion, and limbs. Depending on the conditions, zebras may graze in a fairly small area for a period of time or undertake long daily journeys to graze elsewhere. For example, in the dry season at the Makgadikgadi Pans in Botswana, zebras may walk many miles outward from watering holes—up to 21 miles—to desired foraging areas.[5]

Zebras also migrate from one area to another in the course of a year, often moving to reach areas of rainfall and the period of grass growth.[6] Such migrations are well known in the Serengeti and have been recently recorded in Botswana, with zebras traveling more than 150 miles in one direction to reach a new seasonal home. They may average around 12 to 15 miles per day on journeys that take one to three weeks, but sometimes they travel over 30 miles in a day.[7] That's endurance!

The migrating zebras will pass through areas that offer forage similar in quality to the grassland they are heading for, but they don't stay at them. So there is more to migration than grass availability, and much remains a riddle, including how the animals find their way to such far-off destinations.

When migrating, zebras mostly walk, although they can trot, canter, and gallop like horses, which are close relatives. The faster gaits are typically observed when zebras are fleeing a predator, and they often reach a speed of over 30 mph. They have the stamina to outrun lions if they have an adequate head start. After a longer bout of running, a zebra does not collapse in exhaustion, but remains standing and typically begins to graze again. Running is also a primary play activity amongst zebras. Here a description by Cynthia Moss from her book *Portraits in the Wild*:

The foal may gallop by itself, running around its family and racing up to 150 yards away. Foals may also play racing and chasing games with each other, and even with foals from neighboring groups.... The animals in bachelor groups, especially in those made up of young animals, are very high-spirited and playful. Their running games turn into races, with the whole group of them galloping across the plains at full speed.[8]

Just as it can lower its head to the ground while standing, a zebra can raise its head above the height of its torso. With good senses of hearing (ears are mobile and can be turned in all directions) and smell, and eyes on the side of its head giving it a large field of vision, a zebra can spread its awareness out into its surroundings. A major focus of awareness and interaction is with other zebras in its group or herd. Behavioral ecologist Richard Estes remarks, "Individual members rest, feed, move, groom, dustbathe, suckle, and excrete on much the same schedule, as though these activities were all infectious."[9]

The plains zebra lives in groups with usually one stallion, a number of mares, and their offspring; the mares are not necessarily related.[10] These groups are stable over long periods of time, while the larger aggregations of hundreds or even thousands of zebras—zebra herds—are ephemeral entities that change in composition, and also grow and shrink depending on a variety of conditions.

The lion demands of us a radical shift in perspective to see its way of being in the world. Take grass. While we can say it is the same material

FIGURE 5.2. A lioness stalking. (Drawing after a photo by Peter Blackwell, naturepl.com)

and may in one sense look the same for a zebra and a lion, the meaning of grass in the lives of these two animals could hardly be more different. For a zebra it is food, and much of its daily activity consists in feeding and digesting grasses. A lion watching zebras doesn't suddenly come up with the idea that life would be much more comfortable hunting grass than hunting those zebras. For a lion, grass is a space to rest, hide, and stalk, when long; when short it is an area to saunter, attack, and often to be noticed by prey.

While zebras spend 20 hours a day standing, walking, and grazing, lions spend about the same large portion of the day resting.[11] As Cynthia Moss writes, "Lions are consummate resters; they stretch out in sometimes ridiculous positions, utterly relaxed and apparently without a worry in the world. This quality makes them a difficult animal to study in a way—the scientist is forever in danger of falling asleep with his subjects."[12] Imagine a group of lounging, straw-colored lions on a warm afternoon and you can easily understand why Anne Morrow Lindbergh was moved to say they appear "poured out like honey in the sun."[13]

Such a group of lions will often consist of a few females, their cubs, and perhaps a male or two. They all belong to a larger pride of lions comprised of those lions that peaceably aggregate and interact in a given region. Smaller groupings within the pride change continually, but whole prides have large territories (averaging around 35 square miles) that they defend against other prides.[14] The degree of social interaction within groups of lions is unique among cat species.

As with zebras, play behavior provides a vivid picture of the way lions live in their bodies and of the kinds of movement and attentiveness they express. George Schaller, in his seminal study of lion behavior, describes lion cubs at play:

> The cub paws a twig, then chews it. When another cub passes, he lunges and bites it in the lower back. It turns and swats, then walks away. The cub sits. Suddenly he stalks a cub and rushes. The one attacked rolls over with a snarl and both grapple. The cub desists in its attack and bites at a tuft of grass instead. He then flops on his side. After lying briefly on his back and waving his feet, he rolls over and watches other cubs play. One of these ambles closer. He crouches

behind some grass, then rushes and swats and in the same motion
turns to another cub and nips it in the flank. The other cub whirls
and hits him with a paw. He leaves. Two cubs wrestle, and he grabs
one of these with his paws. One clouts him in the face.[15]

While wrestling is unique to play, other activities such as chasing,
stalking, rushing, pawing, and swatting are all relevant when at a later
age lions start hunting. But in play, everything occurs in a relaxed and
unfettered way.

After a long day of rest, lions often rouse to activity in the late afternoon or early evening. As their drive to hunt awakens, they begin moving through their environment. These beings, which can be so carefree,
loose, and unfocused, can move seamlessly into focused and forward-
striving action. As a lion searches for prey, its senses of sight and hearing
let it reach out into the surroundings and detect what is focal for it—
potential prey. Favored prey animals are fairly large herbivores such as
zebras, wildebeest, and other large antelopes.[16] These animals are generally larger than lions. While an adult male lion can weigh as much as
an adult zebra, lionesses, the primary hunters, weigh only about half as
much as male lions. If preferred herbivores are rare in an area, then lions
will hunt much smaller prey, and in some areas they hunt the large and
powerful Cape buffalo.

When a hunting lion discovers a nearby zebra, it may immediately burst into a sprint and attack.[17] But lions also often stalk. During
daylight, this entails finding tall grasses or other vegetation for con-
cealment, while at night darkness provides ample cover. Most kills are
made at night—a testament to the lion's keen sensory and motor abilities. Hidden in vegetation or in darkness, a crouching lion waits with
all muscles tensed for its prey to approach. Its attention is fully focused
on the prey and its movements. To close in on the prey, the lion may
sneak-crouch toward it.

A sudden thrust of activity follows as the lion sprints for its prey and
lunges for the nape or throat, which it pierces with its canines. The
forelegs grasp the prey and pull it down. The lion may tear into the neck
and rip open the windpipe, so that the animal soon expires. Or, it may
open its gaping jaws, grab and close the prey's muzzle, suffocating it.

Chomping into the flesh and shearing off chunks with jagged cheek teeth, the lion swallows without any chewing. An adult lion may devour from 20 to 50 pounds of meat at one feeding, so to speak of gorging here is to use a descriptive and not a derogatory term. A pride of lions feeding on one zebra can consume the entire carcass in 30 minutes.[18]

A lion cannot outrun a zebra or wildebeest. In its short sprint, it must overtake the accelerating prey. The lion has no endurance in such a sprint. Hunts are often not successful because the prey simply outruns the lion. When lionesses hunt together, they are more likely to have success than when they hunt alone, since they often form a kind of loose circle around an area with prey, and when one lion attacks and the animal flees, it may be unwittingly driven into the area where some of the other lionesses are hidden.

After a successful hunt, the lion usually retreats to a secluded spot and rests, sleeps, and digests for many hours. In the lion we meet a being that lives in extremes between utter relaxation and utterly focused powerful activity.

The Parts Express and Embody the Whole

The unique way of being of an animal expresses itself in its behavior, physiology, and in all its organs and structures. Every part of the zebra expresses zebra; every part of the lion expresses lion. However, it is one thing to "know" that all parts of the animal are expressive of the whole, and another matter to *see* how the whole is at work and embodied in every aspect.

I will focus on skeletal morphology and branch out from there. The skeleton is the most definitively formed structure in an animal. It resists decomposition when the animal dies and can then be studied as a clearly formed memory of the whole animal. In exploring this memory in detail and relating it to other aspects of the animal, the whole can come to life in us.

The Limbs

Like its close relative, the horse, the zebra rests upon long, stable columns of bone (Figures 5.3, 5.4, and 5.5). Born with well-formed limbs, a foal tries to stand right after birth. After about 10 minutes, it can stand;

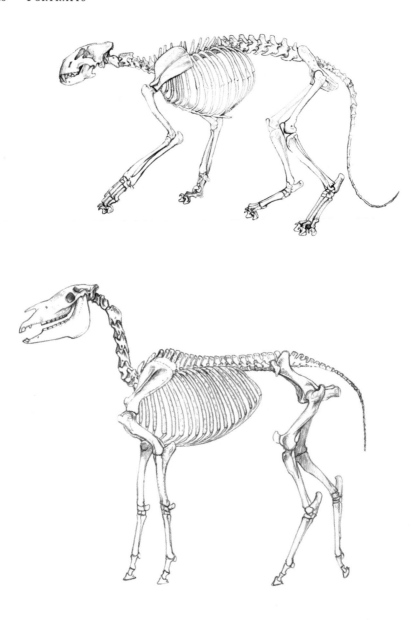

FIGURE 5.3. Zebra and lion skeletons. (Adapted from Kingdon 1977,vol. IIIa, p. 393 and Kingdon 1979, vol. IIIB, p. 144)

it is walking after half an hour; and before the first hour of its postnatal life is over, it is cantering and running about.[19] For an animal to be carrying out its spectrum of movements already an hour after birth shows us that its limbs come ready to act in the world. This is in stark contrast to lions, which are born helpless and kept well hidden. Lion cubs can walk after 10 to 15 days and run around at 25 to 30 days, and they can keep up with the pride after about seven weeks.[20]

The newborn foal has proportionally very long legs and a relatively short body (see Figure 5.1). Over the next few years, the torso doubles in length, while the legs grow only half again as much to reach the adult proportions.[21]

The upper part of a zebra's legs has strong, short muscles that are embedded in the torso. The upper leg bones (which are comparable to our human upper arm and thigh bones) do not extend beyond the torso. The markedly elongated lower parts of the legs below the body (comparable to our lower arms and legs, hands and feet) have only a few muscles. This makes the lower leg of the zebra thin and bony—and relatively light.

Long tendons, which connect muscles to bones, extend from the upper muscles down into the lower legs and feet. In zebras (and horses) some muscles in the lower leg are actually more tendon than muscle and tendons also replace muscle.[22] Tendons are more fibrous than muscle and cannot actively contract. The forefeet and hind feet consist of three bones, and there are no muscles at all in the feet. The foot bones are held together by tough, almost bone-like ligaments.

This structure of bone, ligaments, and tendons provides stability, so that the zebra can stand with virtually no muscular effort.[23] Its muscles can relax in sleep but the zebra does not collapse. In a sense, the zebra's limbs become living architecture. The stability of the zebra limb is connected with the strength of bones, the tight joints, the configuration of muscles, joints, and tendons, and, importantly, the fact that the limb has fewer bones and joints than the limb of other land mammals.

The zebra's forelimbs carry about 70 percent of the body weight. We human beings can rotate our lower arm around its axis; this is only possible because we have two bones—the ulna and radius—that allow this movement. In the zebra, these bones are fused to form one straight, stable bone, which is the longest in the zebra's body (Figure 5.4). Below it a "wrist" is formed by eight thick, compressed bones called carpals. These carpals have horizontal surfaces that rest upon one another and provide stability but little flexibility (Figure 5.5).

The zebra does not stand upon feet with five toes but, rather, upon one enlarged toe that ends in the thick, horned sheath of the hoof. Fewer bones mean fewer joints; the fewer the joints, the fewer the muscles. This all decreases mobility and flexibility, while the stoutness of the remaining bones, along with the fusion of ulna and radius, increases stability. In other words, the flexibility the zebra loses in the leg is compensated for by the stability and strength it gains by becoming living architecture. The zebra can stand, walk, trot, and gallop with great endurance, but cannot crouch to the ground, or easily scratch an ear with its hoof.

The ability to move at high speed for long distances is intimately connected with the structure of the limbs, a connection that has been researched in detail in horses, which have virtually the same body structure as zebras.[24] The hinge-like joints and arrangement of ligaments and tendons allow movement mainly in the forward-backward plane, and at the same time, they restrict side-to-side movement. A relatively small effort of muscle contraction in the upper legs results in a large motion of the long and light lower legs. The elastic tendons in the lower legs and feet act like a spring when the zebra is running, so that with each landing the zebra is propelled ahead. For a zebra, running is hardly more strenuous than walking.

How different is the lion, which tires after a short, forceful sprint, but is capable of such supple and agile movement. This is embodied in its limb structure. There are many joints in the lion's limbs, and the bones are not so tightly connected as in the zebra (see Figures 5.4 and 5.5). The wrist bones, for example, have rounded surfaces—an expression of mobility. They do not possess in themselves the stable architecture of the firmly set, horizontally placed carpals of the zebra. The ulna and radius are two separate bones, allowing rotation of the forelimb by muscles. This rotation comes into play in many of the lion's activities—when it grasps its prey, holds a chunk of a carcass, or cleans itself with its paws. Compared to those of the zebra, the lion's limbs are short and stocky, embedded in an array of muscles.

The lion's front feet have five toes, the back feet four. The body's weight rests on the pads beneath at the base of the outermost toe bones in each foot, extending the lion's characteristic softness and buoyancy far into the periphery of its limbs. Another element of lion movement—the powerful forward thrust culminating in the leap for prey—also comes to expression in the feet, namely in the claws. Held hidden in the paw, they lash out, gash into the prey, and then retract. The activity of the claws in this hunting sequence vividly reveals the way of the lion as a whole—springing forward, penetrating, and withdrawing into inactivity.

Overall, we can say that the lion's stance and movement are directed and modulated at every moment by muscle. By living in the medium of muscle, the lion is capable of both the utmost force and complete relaxation. Moreover, every movement is characterized by a polarity of tension and restraint, power and suppleness.

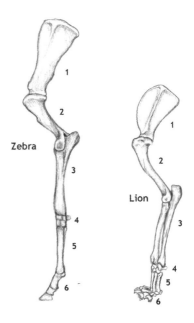

FIGURE 5.4. Forelimb of zebra and lion. 1: scapula; 2: humerus; 3: ulna and radius (in zebra fused together); 4: carpals; 5: metacarpals; 6: phalanges (toe bones).

FIGURE 5.5. Detail of lower forelimb of zebra and lion, viewed from the front.

From Neck to Tail

The zebra has a comparatively long neck that allows freedom of movement in the head. It can lower its head to the ground for grazing, turn its head from side to side, and also raise its head above the level of the body. A typical sight is to see a zebra resting its neck on the back of another zebra. The ridge of the neck supports the mane with its upright hair.

The middle part of the zebra's vertebral column is fairly rigid, although it consists of more vertebrae than the lion's. As in the legs, interlocking surfaces and strong ligaments make the spine a stable, horizontal axis supporting the body through its very structure. This part of the spine keeps essentially the same form under all conditions, whether the zebra is galloping or lying.

The lion's neck, like its limbs, is short and very strong. We need to imagine the neck strength it takes for the lion to pull down a prey animal that it has clamped into with its teeth. In males, the neck (and, of course, the head) is accentuated by the longhaired mane. The middle part of the lion's spine is much more flexible than the zebra's. Its capacity to flex, extend, and bend laterally is much greater. When a lion sprints, the spine rhythmically oscillates between concavity in expansion and arching convexity in contraction. And lying at rest, the lion can stretch out lengthwise or curl up. Because of this flexibility, the body can follow, in its form, any irregularities of the surface upon which the lion lies.

The vertebral column has its continuation in the tail, which is an animal's characteristic extension into the world behind it. The zebra's streaming, longhaired tail emphasizes its vertical aspect. The tail hangs down, is blown by the wind, but also swishes to and fro. The muscular, bony core of the tail extends only into its upper half, while the rest consists of long strands of hair, a substance in which the animal no longer lives. In contrast, the lion lives in its muscular tail almost to the end of its tufted tip. The lion holds its tail actively, not letting it hang down or drag. Perhaps more than any other organ, the tail with its fine undulating movements expresses the lion's momentary state and the inner direction of its alertness. The lion's tail embodies movement and expression, while the zebra's tail, like its lower legs, is more of an organ that is moved.

The Heads of the Zebra and Lion

Through its head, an animal interacts with the world in manifold and characteristic ways. The senses of sight, hearing, smell, and taste are centered in the head, and the sense of touch is most sensitive in the head (tongue, lips, snout). The animal breathes through nose and mouth. And with its jaw—the limb of the head—it feeds.

The zebra has a long head, and just as with the limbs, the part of the head farthest from the body proper is elongated. This front-most portion of the skull is formed by long, tapering bony plates. The snout is not muscular, and its form reveals the underlying bone structure in the same way the lower limbs appear as "skin and bone."

The head's high, broad rear portion is embedded in the neck and jaw muscles. The latter insert into the massive rear section of the lower part of the jaw and connect to the upper, rear part of the skull. The center of gravity is therefore at the rear (body-near) portion of the head, just as large leg muscles are anchored within the body.

The zebra's organs for perceiving what is around and behind it are the nose, eyes, and ears. The eyes are positioned not only sideward, but also quite far back in the skull. The zebra cannot focus on what is directly before it, and it can hardly adjust focus through changing the shape of its lens. In compensation, the zebra can see well in dim light and has a wide field of vision. It sees most sharply what is far away and is sensitive to what moves in the horizontal plane. So while it is grazing with head lowered, it can be aware of what is around and behind it. It has no need to focus on the grass it is feeding on—grass is a field through which it moves, not a point of attention.

So zebras are at home visually in the expanses of the surroundings, which includes awareness of the herd and of predators. Flight from a lion is not movement that involves a focal point; its direction is "away from." By contrast, the lion intently focuses its forward-oriented eyes upon the prey—what is before it and toward which it moves with total absorption.

As figure 5.6 illustrates, the zebra's jaw is dominated by long rows of large cheek teeth. Each row forms one uniform surface that meets with its counterpart in grinding. The image of persistent grinding in which surface meets surface is paralleled by the image of the zebra standing or running with its hard hooves striking the surface of the earth.

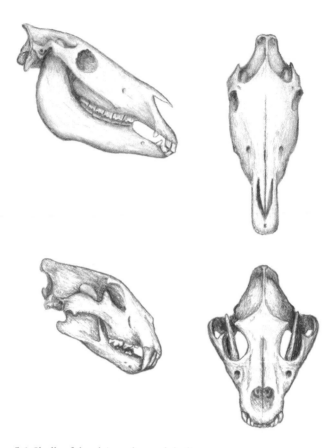

Figures 5.6. Skulls of the plains zebra and the lion.

Like the cheek teeth, the front teeth (incisors) form uniform rows. All these teeth end evenly, so that the zebra can easily clip grass. Characteristically, the one tooth type in mammals that is pointed and never forms surfaces that meet—the canine—is present in the zebra only in a rudimentary form.

The lion's broad, compact skull is almost as wide as the shoulders and hips (see Figures 5.6 and 5.7). The back half of the skull is surrounded by a thick layer of muscles. The space between the broad arcs of the cheekbones and the cranium is filled with the massive jaw muscles. The skull ends in the front in the powerful gesture of the enormous canine teeth.

The lion's canines are as deeply rooted in the upper jaw as they protrude from it. The pointed and conical form of the canines dominates the structure of the other teeth as well. In great contrast to those of the zebra, the lion's incisors are not broad and spatula-like; rather, they are small and have the form of short spikes. The cheek teeth do not form flat surfaces but possess pointed cusps that give them a jagged appearance. When the jaw clamps down (the jaw moves vertically with virtually no lateral movement), the surfaces pass by one another, forming shears that pierce and cut the flesh of the prey.

It is as though the forward thrusting movement of the lion has become frozen in the form of the canines. The claws have a similar form, but they can be extended and retracted.

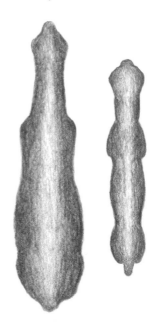

Figure 5.7. Silhouettes of zebra and lion.

There is a further accentuation of this tendency in the frontal positioning of the eyes, which lie quite forward in the short skull. The gaze holds the prey visually before the lion grasps it with claw and tooth. With eyes that are oriented forward, the lion can focus on what is directly in front of it. It can see well in darkness and evidently needs little light to find and kill prey. When there is no moonlight bringing silvery illumination to the savannah, they are more successful in nocturnal hunts.

The male's mane is a majestic image of the forward orientation, which comes more to realization in the activity of the female, who is the dominant hunter.

Anyone who has spent some time where lions live will not forget the experience of hearing roaring lions. Both males and females roar. Roaring typically begins with soft moans, and then a series of "full-throated, thunderous roars,"[25] and finally a series of grunts. This whole sequence can last up to a minute.[26] The deep and loud roars resound far into the surroundings and are often answered by the roars of other lions. The roar grips the whole animal; its jaw is opened wide, and the air streams out and expands the lion far out into the world.

Contrasting Ways of Being

The rhythmically striped zebra attends both to what is close to it—its fellow zebras and the grass it spends countless hours grazing—and to the broad surroundings that may include a distant water hole or its predators.

It lives on the basis of its robust bone structure. Bone is life compressed into solid, enduring form. The leg and foot skeleton has a reduced number of bones, which reduces flexibility, and those that remain grow large and sturdy and in some cases fuse with each other. The stout bones fit tightly together in the joints. The zebra stands upon the ground in the same way that it stably rests upon its own limb bones.

The hooves, which consist of protein, follow the tendency of the robust bone structure—each hoof is a highly thickened and solid toenail that wraps around the last toe bone and provides a solid and stable surface on which the zebra stands and meets the ground when walking or running. The hooves hit the ground when running; the teeth grind grass between their hard surfaces. The zebra meets its world in activity through hard, compact surfaces.

The tendency toward the formation of unified, hard surfaces that we find in the limbs is mirrored in the zebra's jaw, which is its limb in the head. The rows of large cheek teeth are covered with strong enamel, and the zebra spends much of its life grinding highly resistant grass between its teeth.

As it can grind for hours on end, so it stands, walks or runs for most of the 24-hour day. The zebra shows stamina in all its activities—grazing, grinding, digesting, standing, running. Such stamina and constancy is a central aspect of the language of the zebra. Endurance is the physiological and behavioral expression of the stability revealed in the skeletal anatomy.

In contrast to the zebra, the lion lives in its muscles, which function through an interplay between tension and relaxation. The life of a lion oscillates between extremes—focused, powerful action in the hunt, followed by complete relaxation and lassitude.

Stocky limbs and skull express in form the predominance of muscle in the lion. Since its legs are bent when stalking, the lion must draw on enormous body strength to hold its body close to the ground. This con-centrated tonicity erupts into the sheer might of its sinewy being when the lion surges, pulls its prey to the ground, and sinks its teeth into the flesh. The lion feeds ravenously upon the element in which it lives—muscle.

Only an animal that lives to such a degree in the power and tension of muscularity is capable of such complete relaxation. A lion can sleep curled up and entwined with the bodies of its kin, but it could never sleep while standing. It would collapse the moment its muscles relaxed. When the rested lion rises, it stretches every sinew. (Think of your pet cat arching or "concaving" its back.)

Lion activity is not a matter of either force or relaxation separately. In the smoothness, softness, and agile muscular modulation of a moving lion, we can observe the interpenetration of tension and relaxation. The paws roll softly over the ground, and the tail undulates. As seen in the muscular fluidity of body movement, flexibility through joints rather than bony stability predominates in the lion. This quality also manifests in behavior—for example in playful chasing, hitting, and biting between members of a pride. Lion play is a form of "relaxed tension" and reflects the lion's way of life as a whole.

I think we have touched something of the unifying qualities of each of these animals. We can begin to see how different and distinct "parts" can also reveal to us something of the integrated wholeness of the animal. Such wholeness can reveal itself in every aspect of the animal. Whether it does or not depends on our capacity to discern relations between seemingly separate aspects of the animal. Our knowledge will never be "complete"; wondrous riddles will always remain that give a hint at how rich and deep the organic world truly is.

One of those riddles is the beautiful stripes of the zebra—a riddle that warrants its own chapter.

PART 2

Rethinking Development and Evolution

CHAPTER 6

Why Does a Zebra Have Stripes?

(Maybe This Is the Wrong Question)

IT IS HARD NOT TO BE IN AWE of nature's creativity as it expresses itself in the striped coat of zebras.

The rhythmical, flowing sequence of black and white bands of hair is formed as the zebra develops in the womb. Each of the three different species recognized today—plains zebra (*Equus quagga*), Grevy's zebra (*Equus grevyi*), mountain zebra (*Equus zebra*)—has a characteristic striping pattern. And yet there is considerable variation in the pattern in each species and also among individuals. I'll focus on the plains zebra.

A Potent Pattern

It is easy to be mesmerized by the overall impression of the striped animal and fail to perceive consciously what an organic work of art the striping pattern is. We can consider the striping pattern from the perspective of what I will call biological aesthetics: we look closely, moving through the details in such a way that their interrelations and connections with different features of the body and the animal's activity begin

FIGURE 6.1. Plains zebras in the Moremi Game Reserve, Botswana.

Figure 6.2.
Two plains zebras in
the Ngorongoro Crater,
Tanzania. (Photo David
Dennis; Wikimedia
Commons)

to show themselves. The descriptions that follow can only point to what needs to be experienced, so please look at the photos to fill out what the text hints at.

The striping pattern is most complex and refined in the zebra's head, where the senses of sight, hearing, smell, taste, and touch are centered. In the neck and head the animal has greatest freedom of movement—turning down to graze, moving up or from side to side to look and listen.

Between the eyes are long narrow stripes that end in the dark snout; they broaden at the height of the eyes and narrow to the snout and again at the top of the head. The stripes curve around the eyes and the base of the ears. The side of the head has stripes that are perpendicular to the length of the head and curve to converge with the lengthwise stripes to create a wonderfully dynamic pattern.

Overall, there is an interplay of horizontal and vertical striping in the body. Horizontal striping is stronger in the rear of the animal and in lower legs, while vertical striping dominates in the front part of the torso, neck, and head. In front of and above the rear legs, the wide stripes begin on the belly in vertical orientation, then curve toward the horizontal on the rump. As they approach the animal's rear end, each of the black stripes narrows to a tip so that the rear end is more white than black. On the rump the horizontal stripes are broad, becoming

FIGURE 6.3. Plains zebras in the Lake Nakuru National Park, Kenya (Photo Daryona; Wikimedia Commons)

narrower on the legs. The lower part of legs can also have horizontal stripes, which are very narrow. The horizontal striping at the rear covers the rump and the strong leg and pelvic muscles that thrust the animal forward when it moves.

In contrast, the front legs carry most of the body's weight, and the horizontal leg stripes arch upward into the vertical stripes of the shoulder and neck, continuing into the upright standing hair of the mane. It is worth attending to how the flow of the horizontal leg stripes morphs into the vertical stripes of the shoulders. These two streams meet right at the anatomical elbow (which looks like the shoulder) and form a series of upward arching triangular shapes. All this emphasizes the gesture of upward movement.

Each individual animal can be identified by its own unique striping pattern—a whole-body "fingerprint" displayed to the world. You can see this when you look closely at the variation in stripes on the flanks of the animals (see Figures 6.3 and 6.6). A particularly striking example of the individual differences can be seen in the photos of the heads of four

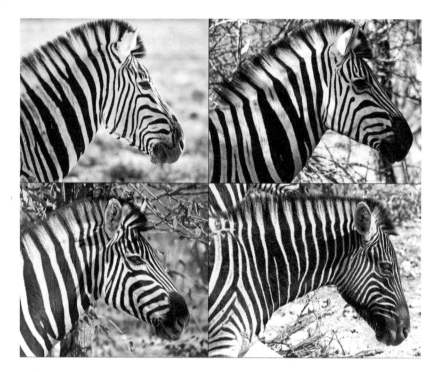

FIGURE 6.4. Variation in the stripe pattern in four different individual plains zebras; Etosha National Park, Nambia. (Photo Hans Hillewaert; Wikimedia Commons)

different zebras in Namibia (Figure 6.4). Such examples show us that we need to be cognizant of the many variations on a theme that occur.

All in all, the rhythmical striping pattern is a harmonious and dynamic whole in which each band relates to its neighbor. Moreover, in the individual variations we witness a kind of creative playfulness of nature that creates sameness (pattern) that is never exactly the same.

Presentation Value

Twentieth century Swiss zoologist Adolf Portmann pointed out that in many closely related species there is more generic sameness in the hidden inner organs and greater species specificity in the external visual appearance.[1] No one would confuse a zebra and a horse based on outer appearance, and, of course, they inhabit different environments. They belong, together with asses, in the genus *Equus*, and their internal organs, skeleton, and muscles are remarkably similar. Only a specialist can tell them apart. Similarly, the lion's skeletal and muscle structure is very similar to the tiger's (both are in the genus *Panthera*), but no one would mistake the striped tiger for a lion.

Portmann coined the term "presentation value" (*Darstellungswert*) to point to the significant attention nature has given to external appearances —how animals present themselves to the world through color, shape, patterns, sound, smell, or texture.[2] Portmann wants us to take the appearances of nature seriously and not to assume they are simply fortuitous results of organic development that "just happen." When we take appearances such as zebra stripes seriously, our awe of nature's creativity grows and at the same time, as Portmann puts it, we are led to a "vista of the inexpressible." In other words, we are confronted with the riddle of what nature is expressing through outer patterns such as zebra stripes. We can try to find connections and relations by comparing patterns, say, in different groups of animals, as Portmann did and more recently biologists such as Wolfgang Schad and Mark Riegner.[3]

It has been noted for a long time that there is geographic variation in the striping patterns of the different subspecies and populations of plains zebras (see Figure 6.5).[4] Generally speaking, in the equatorial region the contrast between the black and white bands is most pronounced, and the stripes extend all the way down the legs (Figure 6.5-1). Farther south, many individuals have lighter stripes between the black and white bands; they are called shadow stripes (Figure 6.5-2 and -3). Then there are the two extinct subspecies of the plains zebra that lived in South Africa. Burchell's zebra (*Equus quagga burchelli*) was more reddish than black, the stripes were not as defined, especially toward the rear of the animal, and they did not extend down the legs (Figure 6.5-4). Finally, the Quagga (*Equus quagga quagga*) had stripes only in the front half of the body and had no leg stripes (Figure 6.5-5).

Of course, within any given population there is, as I mentioned, considerable variation, so that this geographical variation is not linear and clear cut. Biologist Andreas Suchantke noted that on the equator the difference between bright light and shade is stronger than in all other latitudes, given the high daily arc of the sun's path each day.[5] He suggests that the zebra's coat pattern variation in a way parallels changing light relations from the equator to the subtropics, with the shadow-light contrast becoming weaker farther away from the equator. The changing pattern expresses in a surprising way a relation to the changing light environment, without the connection being in any narrow sense adaptive or utilitarian—an intriguing idea. But what gives me pause, as Ruxton[6] points out, is that the mountain zebra (*Equus zebra*) is strongly striped and lives in South Africa, far from the equator, inhabiting areas that formerly only the partially striped Quagga also inhabited.

Asking "What Are Stripes Good For?"

Most professional biologists who have concerned themselves with zebra stripes have asked a narrow question: what are zebra stripes good for? In other words, they "make the implicit assumption," as the authors of a 2015 article about zebra stripes stated, that the stripes are "adaptive."[7] Adaptive means that they must have a specific function that contributes to the survival of the animal. All "appearances," in this view, exist because over time they arose fortuitously through changes in genetic and developmental mechanisms, but were useful to the animal and so were perpetuated through the generations. This is the standard thought form—the conceptual lens—through which biologists today attempt to account for any and all appearances.

The result has been a plethora of stories (often called hypotheses)—at least 18 different ones—to account for zebra stripes:[8] They provide camouflage in tall grass or in poor light conditions; they make zebras look bigger than they are so as to confuse attacking predators; they reinforce social bonding; they help with regulating body temperature; they protect against biting flies such as horseflies or tsetse flies. The list goes on.

In the past couple of decades, some biologists have looked more carefully at these suggested explanations of zebra stripes.[9] They considered the evidence on which the conjectures were based. In some cases there are anecdotal observations supporting the idea of an adaptive function of stripes in specific situations. But more often than not the stories about why zebras have stripes turn out to reflect not any compelling evidence, but rather the researchers' need for some functional explanation.

Tim Caro, a professor of wildlife biology at the University of California, Davis, has done the most thorough examination of "explanations" of

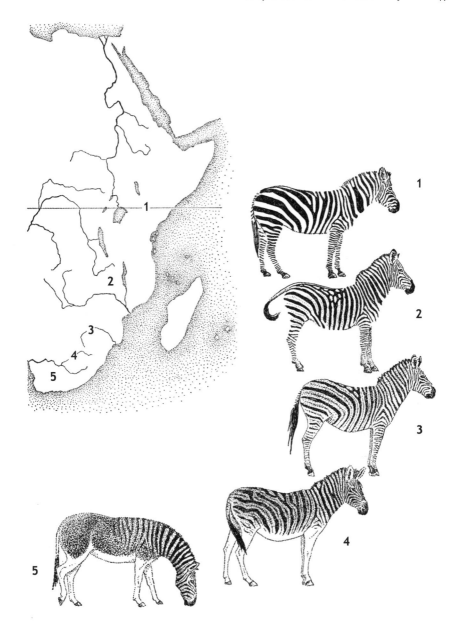

Figure 6.5. Examples of the geographical variation in stripe pattern in the plains zebra (*Equus quagga*). 1: Grant's zebra (*E. q. boehmi*). 2: Selous' zebra (*E. q. selousi*). 3: Chapman's zebra (*E. q. chapmani*). 4. Burchell's zebra (*E. q. burchelli*; extinct). 5: Quagga (*E. q. quagga*). See text for further explanation. (Adapted from Suchantke 2001, p. 8)

stripes. To take one example, he says that "biologists have long remarked on the resemblance between the repeated pattern of stripes on zebras and the vegetation of the habitats in which they live."[10] In tall grass there can be vertical bands of illumination and darkness. And in the early morning and late afternoon there are particularly vivid shafts of shadows contrasting with the brighter vegetation. Some observers have noted in such appearances a certain resemblance to the vertical stripes in the front part of the zebra's torso. Or when a zebra is in woodlands, its pattern can, to a degree, mimic shadow-brightness patterns and also the darker trunks and horizontal branches of trees separated by bands of brightness.

Clearly, such observations of what we could call a kind of "agreement of appearances" can be noted in certain specific conditions as fleeting phenomena, but the agreement (camouflage) is even then very approximate.[11] The zebra does not live in a black-and-white-striped world into which it disappears. And moonless nights, when all animals equally blend with the darkness and from a visual point of view all animals are well camouflaged, are when zebras most often fall prey to lions.[12]

On the whole, zebras are visually highly conspicuous during the daytime, whether in open grasslands or in woodlands. Moreover, "Compared to many hoofed animals on the plains of Africa, they are remarkably mobile and noisy and never attempt to hide in cover."[13] Nonetheless, another popular "explanatory" story is that the striping pattern, especially when zebras are moving and in groups, could in a variety of ways confuse predators. But lions attack and successfully kill zebras whether they are solitary or in groups, and as just mentioned, most kills are at night when stripes could not dazzle or confuse lions. Again, it may be the case that once in a while, in a particular context, the striping pattern confuses an attacking lion for a moment and the zebra escapes. But it is not a general rule.

Caro performed many field experiments to test the hypotheses about stripes and discusses these along with a plethora of other research in his 2016 book *Zebra Stripes*. The careful research summarized in the book leads him to reject virtually every hypothesis. What he does not do is to question whether a survival-based explanation exists. He is not moved to drop his conceptual lens.

Figure 6.6. Two plains zebras grazing during the dry season; Moremi Game Reserve, Botswana.

His own conclusion is that stripes are "an evolutionary response to pressure from biting flies" such as tsetse flies, or horseflies that belong to the Tabinid fly family.[14] Such flies can transmit diseases and cause substantial bleeding in the many large mammals they bite. One supporting observation is that in field experiments some of these flies tend to avoid black-and-white-striped surfaces. Caro and his colleagues believe there is a correlation between biting-fly abundance and the degree of striping in zebras and other members of the horse family (asses, horses). He marshals a number of additional observations and thought connections that he suggests support this hypothesis. In a journal article, he and his colleagues conclude that "a solution to the riddle of zebra stripes, discussed by Wallace and Darwin, is at hand."[15]

Another group of researchers, led by Brenda Larison at UCLA, disagrees. They point out that biting-fly abundance was not directly observed, and there is a lack of data about abundance and distribution. What Caro and colleagues did was to use two environmental conditions—temperature and humidity—as the proxy to estimate the abundance of biting flies, since they tend to be more prevalent in warm, humid conditions. As a result, "What they [Caro et al.] call 'tabanid

distribution' could easily correlate with any number of species distribu-
tions, be they insects, plants, or vertebrates."[16] In their own research on
plains zebras, Larison and colleagues found variation in striping pattern
to be most strongly correlated with temperature. What that correlation
means, and whether—as Larison and colleagues assume—it has to do
with an adaptive function, remains open.

Let's assume, for the sake of argument, that research continues and
researchers were to find that there is a correlation between stronger
striping in zebras in areas with greater abundance of biting flies. And
let's assume, in addition, that widespread field observations of zebra
populations reveal that horseflies are less likely to bother boldly striped
zebras than others. Would we, with this additional evidence, have
"explained" or accounted for zebra stripes? Would we have a "solution
to the riddle of zebra stripes?" Would we know why zebras have stripes?
Of course not.

What we would know is that stripes play a role in defense against
horse flies, just as they may under certain conditions provide camouflage, or make zebras stand out more. But all of these "functions" could
have arisen as propitious side effects of the striping pattern.

Moreover, in such purported explanations, the pattern in its concreteness, in all its nuances and variations, is glossed over because one only
focuses on the abstraction: black and white (or dark and light) surfaces
that are clearly distinguishable. These could be blocks, circles, blotches,
straight bands, etc., so the explanation would not tell us why zebras specifically have *stripes*. The explanation is detached from the real animal.

I do not have an explanation for the zebra's stripes. I am not looking
for an explanation. I'm trying to get closer to what the animal may reveal
as its unique way of being. By attending to the stripes I have been led to
see an expression of nature's creative power, and I am intrigued by each
new variation. The zebra's wonderful stripes remain a riddle for me. I'm
happy to wait and see whether further insights arise.

CHAPTER 7

The Giraffe's Long Neck

From Evolutionary Fable to Whole Organism

A LONE GIRAFFE BULL STOOD AT THE EDGE of a scrubby bush forest that opened into grassland. It was August, the beginning of spring in the southern African savannah, but also the middle of the dry season. The grasses and forbs were yellowed and brittle. Many trees and bushes had no leaves, though some still bore fruit, and others were just beginning to flower.

The giraffe didn't seem bothered by our presence, although we were off the main tourist track. We were quite close, and its towering height was striking. Long narrow legs carried its barrel-shaped, beautifully brown-and-white-patterned body high above the ground. Its back sloped downward, extending into the tail with its long strands of wavy hair nearly reaching the ground. Toward the front the body took on more bulk and sloped steeply upward, merging into the massive, skyward-reaching neck.

From its lofty perch, the giraffe watched us calmly with dark, bulging eyes. It was not excited, nor was it aggressive. When it turned its head to face us directly, we could see fine out-curving eyelashes encircling attentive eyes.

We observed the animal for a good while. It was eating, but not on the leaves of trees and bushes, which we'd grown used to seeing giraffes eat. There were no trees or bushes within its reach, and its head was not lowered to the ground grazing. But it was chewing a hearty meal, part of which was sticking out of its mouth. Imagine a giraffe smoking a giraffe-sized cigar and you get an inkling of the scene. The giraffe's meal was a sausage-tree fruit,[1] which really does look like a sausage (or an over-sized cigar). Sausage trees hang full of them at this time of year. They are about one to two feet long, two to three inches in diameter, and can weigh up to 20 pounds.

About six inches of the fruit were protruding, so the other foot or so was in the giraffe's mouth cavity. Chewing with circling motions of the lower jaw, every now and again the giraffe would raise its head in line with its neck and gulp, as if trying to swallow the fruit. But the fruit never moved. We were concerned that it might be stuck, since, at the time, we didn't know that giraffes do regularly eat these fruits during the dry season. But the animal didn't look concerned and was apparently in no rush; with a sausage fruit as its meal it didn't need to wander around. I don't know how long we were there, but eventually we moved on, wondering whether the giraffe succeeded in getting the long fruit through its long mouth and down into its long throat.

Everything about the giraffe seems built around lengthening—from its tail hairs to its long eyelashes, from its long legs to its long neck and head. Coming across a giraffe embodying elongation to the fullest in eating that long fruit of a sausage tree was an unexpected gift.

* * *

In the 1980s I taught a college-preparatory high school course on evolution in Germany. Although I was teaching at an independent Waldorf school, the curriculum for the final year of high school (thirteenth grade) had to conform to state guidelines. One of the topics involved comparing Lamarck's view of evolution (the wrong one) with Darwin's (the right one). An expedient way to do this was to use the giraffe and its long neck as an example. How would Lamarck and Darwin each explain how the giraffe's neck evolved? Textbooks sometimes discussed this example, so I had the material I needed to introduce, explain, and finish off the problem of Lamarckism versus Darwinism in one three-quarter-hour period, with a review in the next class. Since I was always under time pressure to cover all the material, this example was efficacious and had the added advantage that it stuck in the students' minds.

I don't know how many times I taught this example, but I do remember that both the students and I had a hard time taking it too seriously. It was clear, at least in a subterranean way, that the giraffe was just a handy convenience to make a theoretical point. With knowing smiles, we moved on to more serious matters.

In too willingly following authority ("the curriculum"), I had not taken Lamarckism seriously enough and had hardly given the giraffe the time of day. I was teaching about a caricature—not about the giraffe and not about evolution. How could I possibly teach about an animal's evolution if I knew next to nothing about that animal and was only using its long neck to make a point? When I think about this today, I have to cringe and extend my inner apologies to my students at that time, to Lamarck, and, of course, to the giraffe.

It may be that a need for redemption later led me to become fascinated with the giraffe and to study it in much greater detail. This chapter is a result of that study. In essence, it is a conversation in three parts. I begin with a conversation with evolutionary ideas about the giraffe's long neck. Then I converse with the giraffe itself. And finally the conversation returns to the question of evolution, but from the perspective of the giraffe as an organism.

A Long Neck Reaching Higher

Once scientists began thinking about animals in terms of evolution, the giraffe became a welcome—and seemingly straightforward— example. It is as if the giraffe's long neck were begging to be explained by evolutionary theorists.

One of the first evolutionary thinkers, Jean-Baptiste Lamarck, offered a short description of how the giraffe evolved in his major work, *Philosophie Zoologique*, published in 1809:

> It is interesting to observe the result of habit in the peculiar shape and size of the giraffe: this animal, the tallest of the mammals, is known to live in the interior of Africa in places where the soil is nearly always arid and barren, so that it is obliged to browse on the leaves of trees and to make constant efforts to reach them. From this habit long maintained in all its race, it has resulted that the animal's forelegs have become longer than its hind-legs, and that its neck is lengthened to such a degree that the giraffe, without standing up on its hind-legs, attains a height of six meters.[2]

In Lamarck's view, we must imagine a situation in the past, in which the best food for browsing mammals was higher up in trees, the lower

Figure 7.1. Giraffe in a "classic" feeding position, extending its neck, head, and tongue to reach the leaves of an acacia tree, Kenya. (https://commons.wikimedia. org/wiki/File:Giraffe_Feeds.jpg)

vegetation having been eaten by other animals. The predecessors of the giraffe—which we should imagine to have been like antelopes or deer—needed to adapt their behavior to this changing environment. As Lamarck wrote, "Variations in the environment induce changes in the needs, habits and modes of life of living beings.... These changes give rise to modifications or developments in their organs and the shape of their parts."[3] So Lamarck imagined that over generations the habit of continually reaching for the higher browse produced in the giraffe's ancestors a lengthening of the legs and neck.

A little over 60 years later, Charles Darwin commented on giraffe evolution in the sixth edition of his seminal *Origin of Species*:

> The giraffe, by its lofty stature, much elongated neck, fore-legs, head and tongue, has its whole frame beautifully adapted for browsing on the higher branches of trees. It can thus obtain food beyond the reach of the other Ungulata or hoofed animals inhabiting the same country; and this must be a great advantage to it during dearths.... So under nature with the nascent giraffe the individuals which were the highest browsers, and were able

during dearth to reach even an inch or two above the others,
will often have been preserved; for they will have roamed over
the whole country in search of food…. Those individuals which
had some one part or several parts of their bodies rather more
elongated than usual, would generally have survived. These will
have intercrossed and left offspring, either inheriting the same
bodily peculiarities, or with a tendency to vary again in the same
manner; whilst the individuals, less favoured in the same respects
will have been the most liable to perish…. By this process long-
continued, which exactly corresponds with what I have called
unconscious selection by man, combined no doubt in a most
important manner with the inherited effects of the increased use
of parts, it seems to me almost certain that an ordinary hoofed
quadruped might be converted into a giraffe.[4]

In many respects this is a classic formulation of how Darwin viewed
evolution: every species consists of individuals that show considerable
variations. Under certain environmental conditions particular variations
will be most advantageous. Natural selection weeds out the unadapted,
and the better adapted survive. These variations become dominant in the
species, and so it evolves. In the case of giraffes, times of drought and arid
conditions give an advantage to animals that can outcompete others by
reaching the higher, untouched leaves. They form the ancestral stock of
the animals that evolve into giraffes.

Interestingly, Darwin also believed in the "inherited effects of the
increased use of parts." Evidently, he thought that repeated use by
giraffe ancestors of their somewhat longer necks to reach high vegetation
would increase the likelihood of the longer neck being inherited by
the next generation. Darwin felt this was key to explaining evolution,
since it provides a mechanism for anchoring a novel trait in the heredity
stream. Otherwise a new characteristic might arise in one generation and
disappear in the next. This "Larmarckian" view—that the activity of the
organism affects its evolution—may have made sense to the founder of
Darwinism, but has certainly not been a popular idea among mainstream
Darwinists. However, there has been within recent decades a resurgence
of Lamarckian-like views of evolutionary processes, usually going under
the name "epigenetic inheritance."[5]

An Evolutionary Story Falling Short

The idea that the giraffe got its long neck due to food shortages in the lower reaches of trees seems almost self-evident. The giraffe is taller than all other mammals, can feed where no others can, and therefore has a distinct advantage. To say that the long neck and legs developed in relation to this advantage seems compelling. Why else would the giraffe be so tall? You find this view presented in children's books, in website descriptions of the giraffe, and in textbooks.

But because this explanation is widespread does not mean it is true. In fact, this "self-evident" explanation retains its ability to convince only as long as we do not get too involved in the actual biological and ecological details. So let's look at some of the observations that researchers have made concerning giraffe feeding behavior today.

As browsers, giraffes feed mainly on the leaves of various species of acacia trees and bushes. They feed on leaves that are higher up on trees than those accessible to the many browsing antelopes in the savannah. In an area in Niger where giraffes coexist in an agricultural habitat with cattle and humans, for example, giraffes feed above the height that cattle can reach, and males usually feed at a height of over four meters, while females feed at heights between two meters (neck horizontal) and four meters (neck fairly upright).[6] A study in South Africa showed that trees were more strongly browsed at a low height accessible to multiple browsers than at heights above 2.5 meters, which are accessible mainly only to giraffes.[7] So giraffes do find in general more food at greater heights.

As a rule, giraffes are not usually stretching their necks and heads to reach the highest browse, as a variety of studies show.[8] Males, which are up to a meter taller than females, usually feed at greater heights than females and may feed up to half of the time at heights of almost five meters. The remaining time is spent feeding between shoulder height and top of the neck. Females tend to feed at heights in the range between shoulders and top of neck, and one study even observed that females fed around 70 percent of the time at belly height or below.[9]

And what about during droughts? In East Africa, giraffes have been studied quite extensively.[10] They move seasonally, and in the dry season tend to seek out lower valley bottoms and riverine woodlands. There they usually feed from bushes at or below shoulder height (about two

FIGURE 7.2. Giraffe feeding at about shoulder height—the most prevalent height at which giraffes feed. (https://commons.wikimedia.org/wiki/File:Giraffe_feeding_(5107743448).jpg)

and one-half meters in females and three meters in males). Fifty percent of the time they feed at a height of two meters or less, which overlaps with the feeding zone of larger herbivores such as the gerenuk and the kudu (see Figure 7.2). During the rainy season, when there is abundant browse at all levels, giraffes are more likely to feed from the higher branches, browsing fresh, protein-rich leaves. These observations stand in stark contrast to the idea that during droughts the tallest giraffes survive by feeding on the highest browse.

During a drought in Zimbabwe in 2008, researchers discovered 28 giraffes that had died.[11] Fifteen (54% percent) were mature adults, of which twelve were large males. So a disproportionate number of the biggest and tallest animals died during the drought. The "advantage" of height did not help them. Being so big they did not have enough food—it is estimated that a mature adult giraffe may need up to 110 pounds (50 kg.) of browse each day, while a much smaller one-year-old giraffe may require only 30 pounds (14 kg.) of browse each day. Young adults and adult females were the primary survivors of the drought, and as Graham Mitchell and his colleagues remark, they "were not, as predicted by Darwin's hypothesis, the tallest and most able to reach 'even an inch or two above the others'."

So present day observations don't support the idea that the ability to reach high browse drove the evolution of neck lengthening in giraffes. Just because giraffes have long necks and long legs and *can* reach food high in the trees does not mean that a need to reach high browse was a causative factor in the evolution of those characteristics. Evolutionary theorist Stephan Jay Gould and giraffe researchers Graham Mitchell and John Skinner reach a similar conclusion. Mitchell and Skinner state, "The presumptions of historical unavailability of browse and of browse bottlenecks as the selective pressures for neck and limb elongations are highly doubtful and probably false."[12]

There are, incidentally, other ways to reach, as a mammal, the high foliage of trees. Goats, for example, are known to climb into trees and eat foliage (see Figure 7.3). In Morocco, for example, the diet of goats consisted of 47 to 84 percent leaves of the Argan tree, which the goats must climb to access.[13] Why didn't tree-climbing leaf eaters (folivores) develop in the savannah? The elephant sometimes stands on its back legs and extends its trunk to reach high limbs—but no one thinks that the elephant developed its trunk as a result of selection pressures to reach higher food.

It is hard not to agree with scientists Robert Simmons and Scheepers' comment that the standard explanation of the giraffe's long neck "may be no more than a tall story."[14]

Alternative Stories

After critiquing Darwin's explanation in *Nature* decades ago, biologist Chapman Pincher suggested that the "most extraordinary feature of the giraffe is not the length of the neck but the length of the forelegs."[15] By developing long legs, the giraffe acquired a huge stride so that it can move relatively fast for its size, which left it with only one main predator— the lion. Emphasizing this constellation of facts, Pincher explains the "excessive length of its forelegs as the effect of natural selection acting continually through the hunter-hunted relationship, as in the case of hoofed mammals generally." The neck, in turn, followed the lengthening legs so that the giraffe could still reach the ground to drink.

FIGURE 7.3. Goats do not require a long neck to feed on twigs and leaves of an argan tree (*Argania spinosa*). (Postcard from Morocco)

It is strange that Pincher is able to critique Darwin's view so clearly and yet doesn't recognize that he is proposing the same type of inadequate explanation. The giraffe ancestor could just as well have developed greater bulk or more running muscles, both of which would have aided in avoiding predators. The fact is that despite its size and long stride, the giraffe is still preyed upon by lions. And as one study of 100 giraffes killed by lions in South Africa showed, almost twice as many bulls were killed as cows.[16] The longer stride of bulls evidently doesn't help them avoid lions better than the shorter legged females. Who knows whether their long stride may in some way make them more vulnerable? Another speculative idea into the wastebasket.

Another scientist mentions that the lengthening of the limbs and neck could allow the giraffe to dissipate heat.[17] This would be advantageous in the hot tropical climate since the largest animals would have been best able to survive heat waves, so natural selection would encourage the tendency toward lengthening.

As in the other suggested explanations, the central question is whether the idea is rooted in reality. A key factor in a mammal's ability to dissipate heat is the ratio of its total surface area, through which it can lose heat, to its total heat-producing volume. A small animal has a small body volume in proportion to its surface area, while in a large animal the volume is proportionally large.[18] The giraffe is a very large animal with a barrel-shaped torso. Although its neck is long, it is also voluminous; only the lower parts of the legs, which carry relatively few blood vessels, would act to enlarge the surface-to-volume ratio substantially. But the giraffe's total surface-to-volume ration is not different from other large mammals of similar weight.[19] So despite appearances, the giraffe's long neck and legs do not give it, overall, a larger heat-releasing ability. Giraffes do tend to face the sun on very hot days, which exposes less of their big body to the sun, but this kind of "behavioral cooling" has been observed in other animals with a variety of shapes.

The giraffe's great height and keen eyesight give it the ability to see far around it. As a result it can detect predators, avoid them, and also stay in connection with other giraffes. Could this ability have been the driver of the lengthening of the neck and legs?[20] If this had been so, you would expect, for example, that mothers with calves would be more vigilant in observing than mothers without calves—but there is no evidence that this is the case.[21] Also, one finds fossils of smaller species of giraffes that have essentially the same proportions as today's much larger giraffes, a fact that hardly supports this hypothesis. As two researchers remark, "At the moment there is not a strong logical or empirical reason for expecting that anti-predator vigilance has been an important selector for long necks."[22]

Other scientists propose that sexual selection caused the lengthening and enlarging of the neck in males.[23] These scientists relate their ideas to known facts and point out shortcomings in relation to larger contexts—a happy contrast to the other hypotheses we have discussed. They describe how male giraffes fight by clubbing opponents with their large, massive heads; the neck plays the role of a muscular arm. The largest (longest-necked) males are dominant over other male giraffes and mate more frequently. Therefore, selection works in favor of long necks. This explanation would also account for why males have not only longer, but proportionately heavier heads than females.

This hypothesis seems consistent with the difference between male and female giraffes. At least it gives a picture of how the males' longer neck can be maintained in evolution. But it doesn't tell us anything about the origin of neck lengthening in giraffes per se—the neck has to reach a length of one or two meters to be used as a weapon for clubbing. How did it get that long in the first place? Moreover, the female giraffe is left out of the explanation, and Simmons and Scheepers can only speculate that female neck lengthening somehow followed that of males. In the end, the authors admit that neck lengthening could have had other causes and that head clubbing is a consequence of a long neck and not a cause.

All the above "explanations" of the evolution of the giraffe's long legs and long neck are unsatisfying. Each of the scientists sees problems in other explanations but remains within the same explanatory framework when putting forward his or her own hypothesis. As we saw in all the proposed "explanations" of zebra stripes, no one recognizes the necessity for stepping outside the framework and looking at the difficulties of the overall approach. In each case the scientist abstracts individual features (long neck, long legs, large surface area) and considers them in isolation from the rest of the organism. The individual feature is then placed into relation to *one* purported causal factor in the environment (drought, heat, predator avoidance, male competition). The link of individual feature to environmental factor is supposed to explain the evolution of that feature.

But this procedure is highly problematic. The giraffe's neck carries out a variety of activities that we label as "functions"—it allows feeding from high branches, serves as a weapon in males, brings the head to elevated heights that give the giraffe a large field of view, is used as a pendulum while galloping, and so on.

Virtually all the structures and organs in the animal body are multifunctional and interact dynamically with other multifunctional structures and organs. When we pick out a single function and focus solely on it to explain a multifunctional organ, the explanation can only be inadequate. It is comparable to believing you have accurately portrayed a richly nuanced, multicolored landscape with charcoal. It just does not work.

Does the Giraffe Really Have a Long Neck?

I sometimes wonder why no one has maintained that the giraffe has, in reality, a *short* neck. If you observe giraffes drinking, or, as they occasionally do, grazing close to the ground, then you know what I mean (see Figure 7.4). Giraffes do not drink often, but when they do, they have to either splay their forelegs to the side or bend them severly at the wrist joint. Both procedures take time and are awkward. But only in this way can the giraffe get the tip of its mouth down to the surface of the water. Looked at from the perspective of drinking, the giraffe has a short neck. Antelopes and zebras reach the ground without bending their legs, and the long-legged elephant has its trunk to compensate for its short neck. Only the giraffe and its rainforest relative, the okapi, have necks so short relative to their legs and chest that they must splay or bend their legs.

So why hasn't the giraffe become famous for its manifestly short neck? Why don't we have evolutionary hypotheses explaining how the giraffe got its short neck? The reason is that the giraffe's neck, in other respects or from other perspectives, *is* long. No other mammal has such a long neck in absolute terms or in relation to the length of its torso. We all have seen, in life or in pictures, and been amazed by the standing giraffe, its long neck sailing skyward, compared in comparison to which the ungainly, short-necked drinking giraffe appears a most unfortunate creature.

Whether the neck is long or short depends on our perspective and on the behavior or anatomical context we are focusing on. We understand the giraffe only when we take various perspectives and let it show different aspects of its being. The moment we focus solely on the long neck— and on it solely in terms of a food gathering or some other strategy—we have lost the reality of the giraffe.

Reality is richer than such explanations. The explanation may be in and of itself coherent and logical, but what it explains is not the thing itself but a specter of it—the isolated aspect that has been abstracted from the whole organism. In reality, the organism as a whole evolves; all its parts are multifunctional, facilitating its interactions with its complex, changing environment. If we don't consider all the partial aspects within this larger context, we can have only inadequate explanations devoid of life.

FIGURE 7.4. "Short-necked" giraffes grazing. Giraffes can reach the ground with their mouths to drink or graze only by splaying their front legs (left) or splaying and bending their forelegs (right). (Drawing after a photo in Dagg and Foster 1982)

Another consequence of the usual way of explaining is that the organism itself is atomized into individual characteristics, each having its own explanation. Each part takes on a quasireality of its own, while the whole organism—which brings forth and gives coherence to the parts—degenerates into a kind of epiphenomenon, a mere composite of the surviving parts that "really" count.

So the whole project of explaining the evolution of an animal by abstracting from the whole leads to unsatisfying, speculative ideas on the one hand, and to a conceptual dissolution of the unity of the organism on the other. A more adequate understanding requires that we first investigate the organism as a whole and how its members interrelate and interact within the context of the whole organism and its environment. This holistic understanding can then form the *starting point* for thinking about the evolution of the animal. The famous statement by the evolutionary biologist Dobzhansky that "nothing in biology can be understood except in the light of evolution" is a grand claim, which I believe is, in the end, true.[24] But we have a lot of work to do before we get there, and we should not be satisfied with shortcut evolutionary "explanations."

If evolutionary thought is to have a solid foundation, we must firmly ground it in holistic understanding. As it is, stories of the evolution of traits seem compelling until you look for their context and foundation in the world and discover a pool of quicksand.

FIGURE 7.5. Despite their divergent morphologies, the elk (wapiti), bison, and giraffe all belong to the group of the even-toed, ruminant hoofed mammals. (Adapted from Schad 1977, p. 177)

The Unique Form of the Giraffe

[In Africa] there lives an animal which the Greeks call Camelopardalis, a composite name which describes the double nature of this quadruped. It has the varied coat of a leopard, the shape of a camel and is of a size beyond measure. Its neck is long enough for it to browse in the tops of trees.

This is one of the first written descriptions of the giraffe, penned about 104 BCE by a Greek scholar, Agatarchides.[25] During the Roman era and in Persia and Europe of the Middle Ages, the giraffe was variously described, but always as a composite creature. So Abu Bakr Ibn al-Fuqih around 906 CE: "The Giraffe has the structure of a camel, the head of a stag, hoofs like those of cattle and a tail like a bird."[26]

In 1756, the giraffe received its scientific name *Giraffa camelopardalis.*[27]

Starting around this time and extending through the nineteenth century, natural historians (they weren't yet called scientists) in England, France, and Germany undertook detailed comparative studies of the anatomy and morphology of animals. It was the golden age of comparative anatomy. On the one hand these natural historians wanted to gain an exact picture of the physical structure of every known animal, and, on the other hand, they were interested in patterns and order in nature. Animals may contrast greatly in external shape and appearance but on closer examination reveal similarities in body plan and anatomical structures.

Despite its odd shape and great size, comparative anatomists recognized that the giraffe clearly belonged to the hoofed mammals, the ungulates. They had discovered two main groups of ungulates: the even-toed or cloven-hoofed (Artiodactyla—bovines, pigs and hippos, deer, antelopes, and camels) and the odd-toed (Perissodactyla—horses and zebras, rhinos, and tapirs). Since the giraffe's feet end in two toes, it was easily identifiable as a member of the Artiodactyla and it was found to share even more characteristics with the deer and cattle (bovid) families. To name a few: it has a four-chambered stomach and chews its cud (ruminates); it has horns; and it has no incisors or canine teeth in the upper jaw.

But the giraffe also has characteristics that distinguish it from the deer and cattle families, so it is placed within its own family within the ruminant ungulates. For example, the giraffe's horns are skin covered and lie above the parietal bones, unlike either cattle horns or deer antlers. Another important diagnostic feature is the lower canine, which in giraffes has two lobes, clearly distinguishing it from the single-lobed canine in all members of the deer and cattle families (see Figure 7.6). Unlike the deer and cattle families, which are diverse and species rich, the giraffe family has only two living members—the giraffe and the okapi of the rainforest (which we will learn more about later).

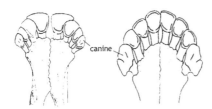

FIGURE 7.6. The lower jaw of the giraffe (right) and the eland (a large African antelope), viewed from above. The large canine (C) in the giraffe has two lobes, a characteristic that only members of the giraffe family possess. (Adapted from Grassé 1955, p. 661)

Knowing that the giraffe is a ruminant, even-toed ungulate provides one starting point for understanding it better. But this knowledge has to become more than simply fitting the giraffe into an abstract biological scheme.

Qualitatively, the more you know about the characteristics of the giraffe and of other ungulates—antelopes, zebras, buffalo, etc.—and learn to view the characteristics in relation to each other, the more the different animals begin to illuminate one another. The unique characteristics of the giraffe speak all the more strongly when viewed in the light of their zoological relatives. The long (or short) neck, the long legs, the large eyes, the particular gait—all become expressions of the unique creature we are trying to understand.

From the perspective of comparative anatomy, the giraffe is in no way a "composite" creature. Like every other living being, it has its own integrity, but the giraffe is also part of a larger web of relations that allow us to understand it better.

In the giraffe we find the characteristics of ruminant ungulates evolved in a singular fashion. It is this singularity within its broader contexts that I hope to illuminate.

Soaring Upward

Charles Darwin identified a key to open up a holistic understanding of the giraffe when he remarked on its "lofty stature, much elongated forelegs, head, and tongue." [28] It is not only the giraffe's neck that is long. We find remarkable elongation in other features.

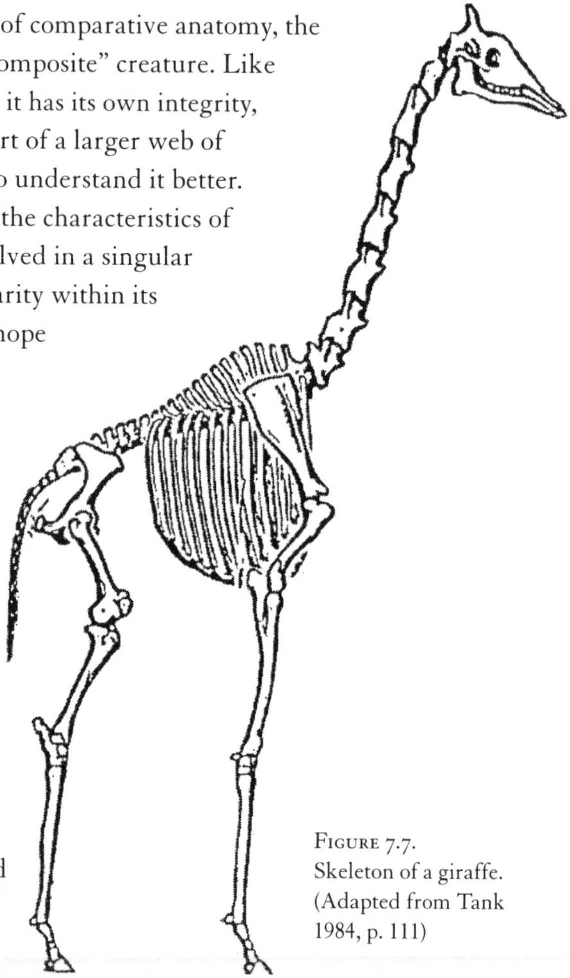

FIGURE 7.7.
Skeleton of a giraffe.
(Adapted from Tank
1984, p. 111)

The giraffe has the longest legs of any animal that is alive today. The ungulates are typically (but not always — think of pigs and hippos) long-legged mammals. This elongation arises primarily through the lengthening of the lower (distal) part of the fore- and hind legs (see Figures 7.7 and 7.8).

Compared with the human skeleton, the ungulates stand on their tiptoes, which are covered with a hoof (thickened toenail). The bones of the toes and feet are very long so that the heel (hind legs) and wrist (forelegs) are high off the ground.

The lower leg bones (ulna and radius in the front; tibia and fibula in the back) are usually fused and long. The elbow (front) and knee (back) joints are high up, so that the relatively short upper (proximal) leg bones (humerus in the front, femur in the back) are taken up into the body.

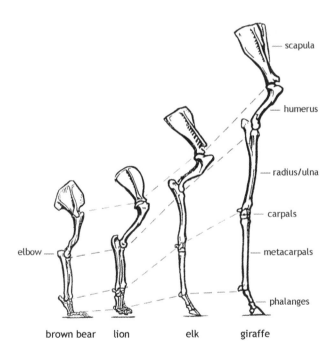

FIGURE 7.8. The foreleg of different mammals; the humerus in each animal has been drawn to the same length. This comparison shows that the lengthening of the giraffe's foreleg is most pronounced in the lower (distal) segments of the leg: radius/ulna and metacarpals. (Original figure; drawings of individual legs adapted from Tank 1984)

The giraffe takes the lengthening of the leg bones to an extreme. The adult giraffe's heel, for example, can be one meter above the ground. The impression of lengthening is emphasized even more because the form of the upper part of the leg remains visible, which is not the case in most other ungulates. In zebras, deer, and antelopes the upper part of the leg is hidden from view by a flap of skin that reaches from the knees and elbows to the torso. The upper leg appears to be part of the torso. Since the giraffe does not have this flap of skin (neither do the relatively long-legged and long-necked camels and llamas), the upper leg is clearly visible.

The giraffe's leg bones are not only long, but also straighter and more slender than those of other ungulates, increasing the character of upright lengthening (see Figure 7.9).[29] Normally in mammals, when the limb bones lengthen they also become proportionally larger in diameter. But this is not the case in the giraffe. The diameter of the limb bones is smaller than it "should" be. The giraffe compensates for its slender bones by making them sturdier and reducing the diameter of the marrow cavity (see Figure 7.10).

In all ungulate species except the giraffe, the rear legs are longer than the forelegs. The giraffe, in contrast, has slightly longer forelegs than hind legs.[30] The length of the forelegs is accentuated by the fact that the humerus rests fairly upright on the vertically oriented lower leg (the fused ulna and radius). In other ungulates the humerus is more horizontally inclined. This vertical orientation is continued into the very long and narrow shoulder blade (scapula).

Similarly, the hind leg—though not as long—has a fairly upright femur. Although the giraffe's pelvis is relatively short, it is, characteristically, oriented more vertically than in other ungulates. Thus the limbs of the giraffe are not only longer in absolute terms, but the bones' vertical orientation increases their upward reach even more.

Viewed from this perspective, the neck follows the same principle as the legs—upright lengthening.[31] It is an astounding fact that all mammals (with just two exceptions, the sloth and the manatee) have seven neck (cervical) vertebrae. Birds with long or short necks have varying numbers of neck vertebrae, but not mammals. Whether the neck is very short (dolphins) or long (giraffe), there are seven cervical

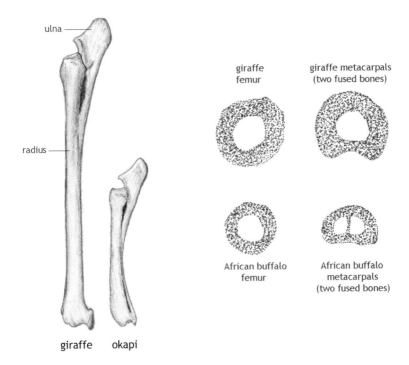

ulna

radius

giraffe   okapi

giraffe
femur

giraffe metacarpals
(two fused bones)

African buffalo
femur

African buffalo
metacarpals
(two fused bones)

FIGURE 7.9. (left)The ulna and radius in the giraffe and okapi, drawn to scale. These two bones are part of the lower leg and are fused together. Functionally they form one stable bone. The giraffe's radius is especially straight and sleek in form. Total length of both bones: 93 cm in the giraffe and 50 cm in the okapi.

FIGURE 7.10. (right) Cross sections of the femur and metacarpals in the giraffe and the African buffalo (*Syncerus caffer*). Both animals weigh about the same. See text for further description. (Drawing after photo in van Schalkwyk et al. 2004, p. 313)

vertebrae. In the virtually neckless dolphin, the seven neck vertebrae have fused to make one short bone that links the head to the torso. Neck lengthening in mammals is achieved through lengthening of the individual neck vertebrae. This is, again, taken to an extreme in the giraffe. In the adult male an individual neck vertebra can be over 12 inches (30 cm) long! Although its limbs are already so long, the giraffe has a significantly longer neck in relation to its limbs than other mammals.[32]

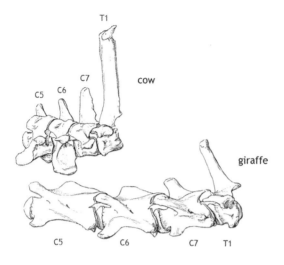

FIGURE 7.11. (left) The last three neck (C5, C6, C7) and first three thoracic or rib-carrying vertebrae (T1) in the domestic cow and the giraffe. The giraffe's neck (cervical) vertebrae are both long and uniform in shape. (From Lankester 1908, pp. 321 & 323)

In comparison to other ungulates, the giraffe's neck vertebrae are not only longer, but also more uniform in shape (see Figures 7.11 and 7.12).[33] The sixth neck vertebra in other ungulates has a unique shape that sets it apart from the other neck vertebrae. But, as anatomist Ray Lankester observed over a century ago, in the giraffe it is very similar to the third, fourth, and fifth neck vertebrae. The seventh neck vertebra in the giraffe is also similar to the rest, while in other ungulates the seventh cervical has become like the first thoracic (rib-carrying) vertebra. Moreover, the first true thoracic vertebra in the giraffe articulates with the vertebrae in front of and behind it in the manner of a *neck* and not a thoracic vertebra. The shape of the first rib-carrying vertebra in the juvenile giraffe is virtually identical to that of the seventh neck vertebra in the adult or juvenile okapi. In addition, the confluence of nerves that serve the shoulder and foreleg (brachial plexus) forms around the first rib-carrying vertebra in the giraffe and around the seventh neck vertebra in the okapi.[34]

These seemingly esoteric anatomical details are eminently revealing. The first rib-carrying vertebra in the giraffe has, in effect, become an eighth neck vertebra. In other words, the tendency to form neck vertebrae extends down into the torso of the giraffe, while in other ungulates the tendency to form thoracic vertebrae extends up into the neck. In the giraffe, the neck has truly become a dominant element in the formation of the spine.

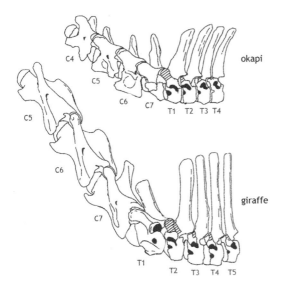

FIGURE 7.12. (right) The last neck and first thoracic vertebrae of the okapi and giraffe. In the giraffe, the first thoracic vertebrae (T1) has essentially become part of the neck. (Adapted from Solounias 1999, p. 264)

This dominance is also visible in the "truly extraordinary" development (as the eminent nineteenth century comparative anatomist Richard Owen put it) of the primary ligament of the neck, the ligamentum nuchae.[35] In mammals, this elastic ligament extends from the head over the neck to the neural processes of the vertebrae between the shoulders. It helps support the head and neck, anchoring them to the torso, and at the same time its elasticity allows movement of the head and neck. It is most developed in longer-necked mammals, so it is not surprising that it becomes the dominant spinal ligament in the giraffe. What astounded the comparative anatomists of the nineteenth century, who examined it for the first time, was that the ligament in giraffes also has muscle-type striated fibers that allow active contraction (as opposed to the fibers in other ligaments, which contract only in response to previous stretching). Describing this ligament, anatomist J. Murie allows himself to express his awe in an otherwise quite dry anatomical study:

> For several reasons this most remarkable body of contractile tissue has been looked upon with the eye of wonder as well as curiosity. Its immense length, volume, and resiliency give it a conspicuous character, added to which it is unique in the ultimate fibre being striated.[36]

Since the giraffe's enormous neck is involved in every movement of the body, this unique ligament that is also a muscle plays a key role in the giraffe's ability to finely adjust the position and movement of its neck.

The dominance of the neck becomes drastically apparent when one compares the length of the neck to the length of the body (thoracic and lumbar part of the spine). The giraffe's neck is 129 percent of the length of the body, while in the white-tailed deer it is 48 percent and in the horse 52 percent.[37] This extreme difference is related not only to the lengthening of the giraffe's neck, but also to the shortening of its body. The giraffe's sacrum—the last section of the vertebral column before the tail—is, for example, very small and short.

In other long-necked mammals, such as the horse and zebra, the camel, the llama, and the gerenuk (a very long-necked African antelope), the body is relatively long and horizontally oriented. These mammals also have relatively long legs. Only in the giraffe, where neck and leg elongation is taken so far, do we find a correlative *shortening* of the body. As Goethe, the scientist who coined the term "morphology," suggested,

> We will find that the many varieties of form arise because one part or the other outweighs the rest in importance. Thus, for example, the neck and extremities are favored in the giraffe at the expense of the body, but the reverse is the case in the mole.... Nothing can be added to one part without subtracting from another and vice versa.[38]

The shortness of the giraffe's body is accentuated by its diagonal orientation. The neck seems to continue down into and through the body into the legs, whereas in other longer-necked ungulates the horizontal orientation of the body is clearly demarcated. In the giraffe, the body appears as a continuation of the neck. The contrast between the stringently horizontal spine of an elk (wapiti) and the upward sloping spine of the giraffe vividly illustrates this difference (see Figure 7.13).

We can learn a great deal about formative tendencies and how different characteristics are interconnected in shaping an animal by carefully comparing the skeletons of the bison, the elk, and the giraffe shown in Figure 7.13. The bison has short legs and a short neck that slopes downward. The elk has longer legs and a longer neck that allows the head to rise above the height of the body and also to reach the ground.

FIGURE 7.13. Skeletons of the elk (wapiti), bison, and giraffe, drawn to scale. (Adapted from Tank 1984 and Olsen 1960)

The bison's limb bones are about the same length as its body, while the elk's limbs are 20 (foreleg) to 50 (hind leg) percent longer than the body.[39] The giraffe's legs are more than twice the length of its body and its neck no longer mediates easily between reaching up and down as in the elk. Its head remains primarily at or above the height of the shoulders, just as the bison's head remains below shoulder height.

The bison's head is held low to the ground, and its rib cage also slopes toward the ground. The neck is attached to the body just above the level of the joint between the humerus (upper leg bone) and the shoulder blade. In the elk, with its horizontal spine, the neck attaches higher up, at about the middle of the shoulder blade. Finally, in the giraffe, the neck

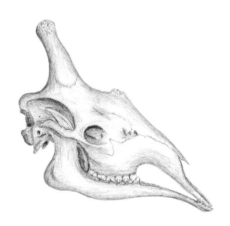

Figure 7.14. Side view of the skull of an adult giraffe from Botswana. Probably a male because of its large size (length of skull: 72 cm [29 inches]) and the massive horns. (Specimen #24290 from the American Museum of Natural History, New York)

attaches at the top of the long shoulder blade. From this already substantial height, the neck soars upward. In other words, the giraffe's neck gains increased height by its high origin in the body.

This relation between neck and body sheds light on why, from the perspective of reaching the ground and drinking, the giraffe has a *short* neck. If its forelimbs were shorter and its neck originated lower down, say, closer to the top of the humerus, it would be able to reach water without splaying its legs. It could drink straight-legged like other "reasonable" mammals. But then it would no longer be a giraffe! You cannot have a towering neck and expect it to reach the ground as well. It is the configuration of the whole body that gives the giraffe such a great height, to which its long neck contributes. But this very configuration makes the giraffe's neck *short* when it returns from the heights to make contact with the earth and water.

The head also reveals a tendency toward lengthening. A male giraffe's skull can be about two feet (60 cm) long. The jaws are long and slender; the front part is especially elongated, and there is a large gap (diastema) between the molars and the incisors (see Figure 7.14). Just as the legs lengthen primarily in the endmost (distal) members, so also does the skull. In his study of giraffe anatomy, Richard Owen remarked on "the prolongation and extensibility of the hair-clad muzzle.... The form of the mouth of the Giraffe differs from every other ruminant ... in the elegant tapering of the muzzle."[40]

The joint between the skull and the neck is also uniquely formed in the giraffe, allowing it to extend its head upward in line with the neck. In this way the giraffe reaches even greater vertical heights.

The tongue puts the finishing touch on lengthening. It is long, slender, and flexible and can extend 16–20 inches (40–50 cm) *beyond* the mouth. So when we hear that a giraffe can be up to 20 feet (6 m) in height, we must remember that it can expand this length another yard or so by raising its head and extending its tongue.

When we say that the giraffe has a long neck, and perhaps add that its legs are also long, we are not saying anything false. But such statements suggest that everything else in the giraffe is "normal" (i.e., typical ungulate) and the long neck is just a matter of extension. In this vein well-known biologist Richard Dawkins writes, "If an okapi mutated to produce a giraffe's neck it would be ... a stretching of an existing complexity, not an introduction of a new complexity."[41]

Dawkins is arguing that one could imagine, fairly simply, the evolution of an okapi-like animal into a giraffe, since this would be just a matter of altering a preexisting organ. But what he overlooks is that the configuration of the whole animal—in relation to the long neck—is altered. You cannot simply get a giraffe by elongating an ungulate ancestor's neck. The whole animal is differently configured. The "story" is elegantly simple, but has little to do with reality. Long legs and long neck are only the two most glaring instances of what we discover to be an overall formative principle in the giraffe—vertical lengthening, which we also see in the head, tongue, and the way the neck attaches to the body. This tendency

FIGURE 7.15. A captive giraffe reaches with neck, head, and tongue to gather browse. (Photo by Mark Riegner, used with permission)

correlates with a shortening in the horizontal (short body and pelvis)—
and even these shortened parts become more vertically oriented than in
other ungulates. The giraffe soars upward.

Mediating Extremes: The Giraffe's Circulatory System

When you survey the scientific literature on giraffe biology, you come
across a large number of articles on its circulatory system. Scientists have
long thought that the giraffe's large body, long legs, and long neck must
place special demands on its circulatory system. As the title of one article
asks, "How does the giraffe adapt to its unique shape?"[42]

One phenomenon that has long fascinated scientists is the giraffe's high
blood pressure—at the level of the heart nearly twice that of most other
mammals.[43] This pressure is continuously and actively modulated and
maintained by the rhythmical contractions of the heart's muscular walls
(especially the thick-walled left ventricle) and other large arteries below
the level of the heart. The primary internal sense organ for perceiving
changes in blood pressure in mammals (the carotid sinus) is located above
the heart, in the carotid artery of the neck, which brings most of the blood
to the head. The giraffe's carotid sinus is highly elastic and finely innervated with sympathetic nerve fibers.[44] We must imagine that through this
internal sense organ the giraffe can perceive blood pressure fluctuations
and then, through changes in heart rate, dilation of vessels, etc., finely
modulate its blood pressure in the whole circulatory system.

There are large pressure differences in different parts of the giraffe's
body. The pressure is, on average, remarkably high in the giraffe's lower
legs and near average (for mammals) in the head and upper neck. One
primary factor in this pressure gradient between head and limbs is the
giraffe's great height and the effects of gravity that come with having
such a large body. Think for a moment of the giraffe as a column of fluid:
the pressure in the legs would be much higher than the pressure in the
head, just by virtue of the weight of the fluid itself. (Imagine the pressure
we feel in our ears at the bottom of a 10-foot-deep pool, where we have
a 10-foot-high column of water resting on us. The higher the column is,
the greater the pressure, which is called gravitational or hydrostatic pressure.) An animal has to deal with such physical forces. And because of its
great vertical length, hydrostatic forces play a more significant role in the

giraffe's life than they do, say, in the life of small animals or even large animals that are essentially horizontally oriented.

But the giraffe is not a static column of fluid; it is a living, active being. Its blood courses through the body, and the body itself changes its positions (standing with raised neck, neck lowering, lying, etc.) and moves at varying tempos. When the giraffe moves, its blood pressure fluctuates radically. For example, each time its forefeet hit the ground while running, the arterial blood pressure (measured in the neck) drops rapidly.[45] Even during normal walking, blood pressure in the feet varies significantly.[46] Evidently, the dynamics of the giraffe's circulation are intimately related both to its overall vertical length and to the movement of its limbs.

So what happens when the giraffe brings its head down to ground level to feed or drink? We know that pressure in the neck artery rises significantly, but backflow of blood to the head through the jugular vein is prohibited by valves that close when the head sinks below body level. But how the giraffe prohibits too much arterial blood from rushing into its brain and then maintains constant blood flow to the brain when again it lifts its head four meters high after drinking, remains a riddle. Biologists have long wondered why a giraffe doesn't get dizzy or faint when it lifts its head. (Think of our tendency to black out when we get up rapidly after lying down.)

There is some evidence that vessels below the head constrict when the giraffe raises its head, keeping blood from dropping away.[47] But one thing is sure: giraffes lift their heads rapidly and show no signs of dizziness or fainting. They maintain their quiet, attentive demeanor.

The center of the circulatory system, the giraffe's heart, lies high above the ground (about 2 meters). Only the elephant has an equally elevated heart. But, in contrast to the elephant, the giraffe's blood courses from the heart another two meters upward through the long neck vessels to the head and back again. The giraffe's heart, therefore, mediates a much larger vertical span than in other mammals, whose dominant orientation of blood flow is horizontal. In the giraffe, as we have come to expect, vertical orientation predominates. This is even mirrored in the shape of the heart itself, which can be over two feet long (!) and is positioned fairly upright within the chest cavity. In contrast, the massive heart of the elephant is much more compact and nearly as broad as long, indicative of its compact body.

neck artery (carotid)

lower leg artery

FIGURE 7.16. Cross sections of a neck artery and a lower leg artery in the giraffe. Both arteries have the same outer diameter. (From Goetz and Keen 1957, p. 552)

For a long time it was thought that the giraffe's heart is particularly large in relation to its body size, but this is not the case. As in most mammals, the giraffe's heart weighs about 0.5 percent of overall body weight.[48] So in a giraffe that weighs a ton (1000 kg.), the heart weighs around 11 pounds (5 kg). The left ventricle in an adult giraffe, however, is proportionally enlarged and has thick muscular walls. The walls thicken as the giraffe grows, its neck lengthens, and its blood pressure increases.

The special nature of the giraffe's circulatory system is further revealed in morphological and functional differences found above and below the heart. Its lower leg arteries have tiny openings and proportionally very thick muscular walls (see Figure 7.16). We must imagine that these vessels are continuously counteracting high gravitational pressure, with their muscular walls acting as a kind of "limb heart" that modulates pressures in the legs. This effect is increased by the tight skin of the legs that helps prevent swelling (edema), which would occur were the vessels thin-walled and embedded in a loose, expandable matrix.

Above the heart, different relations reign. In contrast to the leg arteries, the neck arteries, such as the carotid artery, have wide openings, and are thin walled and elastic. While the leg vessels carry fairly small amounts of blood—the lower legs and feet are virtually skin, tendons, and bones— the neck arteries bring large amounts of blood to the head. They also do not actively contract, as do the leg arteries; rather, they take in the pulsations created by the heart and gradually even them out (the so-called windkessel effect). Blood flow and pressure even out even more when the blood spreads into a network of tiny blood vessels below the brain called the carotid rete mirabile. As a result, the arterial blood flow to the head is much steadier than the pulsating flow in the lower part of the body.

The brain needs—and receives—a constant, steady, and even flow of blood despite changing conditions. The radical fluctuations observed in blood flow and pressure in the lower legs is unthinkable in the brain, which can easily be damaged when, through injury or disease, it receives too little blood even for a short time. From a physiological perspective, the brain is the organ that needs the greatest degree of constancy. (It swims, for example, almost weightless in cerebrospinal fluid and thus is not subject to gravitational forces.) Although the giraffe's head is perched so much higher above the heart than in other mammals, the average arterial pressure in the neck beneath the brain is about the same as in other mammals. Evidently, this pressure is needed to help maintain constant blood flow to the brain.

The contrast in the circulatory system between the neck and head on the one hand and the legs on the other is mirrored in the bony structures of these body parts. The giraffe's leg bones are the densest bones in any four-legged animal, dealing the most with gravity in carrying the animal's weight. Carrying the body so high above the ground, the giraffe's long, sleek legs are subject to special strains. Just as the leg arteries have small openings and thick walls, the leg bones gain strength by laying down extra bone and reducing the size of the marrow cavity within the bones. This makes the bones stable, but they remain sleek, since the overall diameter remains small. In contrast, the long neck vertebrae are much less dense. A kind of expansive lightness extends into the elevated head: the weight of the giraffe's skull is about half that of the skull of a cow with nearly the same body weight.[49] Through its lofty head with its senses of sight and hearing, the giraffe opens itself to the broader surroundings that its long and sturdy legs carry it through.

The Giraffe in its World

There is nothing like seeing a giraffe in its natural habitat—dry savannah grassland with groups and thickets of thorny bushes and trees. When a giraffe stands in or moves across an open grassland, you can see it from far away. It is conspicuous like no other animal. After spotting an individual or group of giraffes when I was observing them in Botswana, I would take my binoculars to view more closely. Invariably I found the giraffes already looking at me (or at least at the Land Rover I was perched in). The giraffe has the largest eyes among land mammals.[50] Since its eyes are set toward the sides of a head that rises four to five meters above the ground, the giraffe has a very large field of view. It is keenly aware of moving objects in its visual field. In viewing the giraffe from afar, you have the impression of a lofty creature sensitive to the happenings within its broad horizon.

When you leave the open grassland and wind your way slowly through wooded and bush areas, you often encounter giraffes, without any preparation, in close proximity. Among trees, the giraffe seems to disappear into its habitat—a stark contrast to its visibility in the open landscape. At least two features of its appearance allow it to blend in this way. First, with its long upright legs from which the neck branches off at an angle, the giraffe's form follows the lines of the tree trunks. When as observers we are close to the ground looking horizontally, what we see (or rather overlook until it's very close) are the narrow legs that meld in among the many trunks of the acacia or mopani trees. The second factor is the giraffe's uniquely patterned coat. Despite the variety of coat patterns in different populations and subspecies of giraffes, all have in common the brown (varying from reddish to black) patches separated by white spaces or lines. When a giraffe is among trees, this dark-light pattern is similar to the mottled light pattern that plays among the branches and leaves. So with its unique shape and coat pattern, the large giraffe recedes into its wooded environment.

It is also the case that the giraffe does not make much noise, either while feeding (browsing the trees and bushes) or after it notices you. It may stand and watch you from on high for a moment, swing its head and neck around, and then amble off. Rarely it may make a snorting sound during such encounters, but that is usually the limit of its minimal aggressiveness. In contrast, an elephant may tread silently, but it loudly breaks off branches while feeding, and trumpets loudly and makes a mock charge when surprised.

Sensing

With its "lofty stature" (Darwin), the giraffe commands a large overview. It's not surprising that the sense of sight plays a dominant role in the giraffe's life. It can see fellow giraffes, and also predators such as lions, from far away. The giraffe's vision is keen—as already mentioned, a giraffe usually sees you before you see it. Experiments on giraffes in captivity indicate that they also see colors.[51]

As we might expect, vision plays an important role in communication between giraffes:

> Staring seems to be a favorite form of giraffe communication. There are what look to human observers like hostile stares, come-hither stares, go-away stares, there's-an-enemy stares. When giraffes spot lions in the grass, a steadfast gaze alerts dozens of other giraffes that may be scattered over a square mile of savanna. Giraffe mothers stare at other adults to warn them away from calves.[52]

The dominant role of vision goes hand-in-hand with a reduction in importance of the sense of smell, so central in most other mammals:

> The sense of smell recedes in importance and is limited to scents in rising air currents.... The unique body of the giraffe causes the sense of smell to play such a small role. Scent-marking of territory

FIGURE 7.17. A lone giraffe walks across an opening in the savannah of Botswana.

falls away …[and] scent glands are lacking. Extensive visual communication compensates the lack of olfactory communication. Tail movements serve as signals.[53]

With its body high off the ground and the head resting even farther up on the long neck, the giraffe distances itself from the rich world of smells near the ground, a world in which most other mammals are immersed. It is a telling fact that the end of the giraffe's nose and muzzle is dry, in contrast to the moist nose and muzzle of most other ruminants.

It is not so clear how significant the sense of hearing is for giraffes. Giraffes themselves make few sounds—snorts, sneezes, grunts and the like.[54] Observations in zoos show that when some giraffes are kept inside and out of view of other giraffes that are outside, they can communicate with infrasound. The visible cue for a human observer is when the giraffes throw back the head and neck, extending the head upright. At this moment they create the deep tones, and immediately thereafter the giraffes outside react. Sometimes the head and neck throws were also accompanied by "a 'shiver' or vibration extending from the chest up the entire length of the trachea."[55] Unlike with elephants, it is not yet known whether and how they might use this capacity in the wild.

Floating over the Plains

One of the most striking things about giraffes is the way they move. Karen Blixen describes the strange quality of their movement beautifully in *Out of Africa*: "I had time after time watched the progression across the plain of the giraffe, in their queer, inimitable, vegetative gracefulness, as if it were not a herd of animals but a family of rare, long-stemmed, speckled gigantic flowers slowly advancing."

An adult giraffe can weigh over a ton, yet its movement appears almost weightless. The giraffe has two different gaits—the ambling walk and the gallop. In contrast to most ungulates, the giraffe walks by swinging its long legs forward, first both legs on one side of the body and then both legs on the opposite side. This type of stride is called an amble, and the giraffe has it in common with okapis, camels, and llamas. In contrast, other ungulates walk by simultaneously moving the left front and right rear legs and then the right front and left rear legs. The amble has a flowing, rhythmical quality, and the giraffe's body and neck softly swing side to side, counterbalancing the one-sided movement of the legs.

Since the giraffe's legs are longer than any other mammal's, it also has a very long stride. Also, with the forelegs being longer than its hind legs and its short body, its gait is unlike that of any other mammal. Its rear leg touches the ground about 20 inches (50 cm) in front of the spot from which it lifted its front leg. Because the giraffe is so large, the movement of the legs seems to be almost in slow motion. And with its center of gravity so high, and its attentiveness concentrated in the elevated head, the giraffe seems to sweep along, hardly in contact with the earth. It treads on the earth, but it certainly does not appear to be of the earth. As Jane Stevens describes, "I watched as a group of seventeen floated along the edge of a yellow-barked acacia forest."[56]

The unearthly quality of movement intensifies when the giraffe accelerates into a gallop (see Figure 7.18). Its stride lengthens even more and when its fore- and hind legs are widely spread and the forelegs reach far forward, the neck becomes more horizontal. The feet then come close together and at this phase of the gallop the neck reaches its most vertical position. The faster the giraffe moves, the more its neck moves down (forward) and up (back). A giraffe can attain a speed of about 35 to 45mph (55–65kmh). The long swinging movements of both the legs and neck and the rhythmical expansion and contraction (spreading out

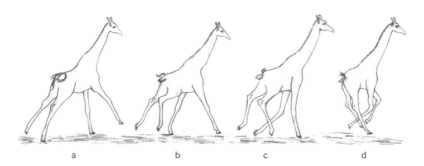

a   b   c   d

FIGURE 7.18. A galloping giraffe. a): The most extended phase of the gallop as the left foreleg has reached the ground. b): The right foreleg reaches the ground. c): The right foreleg is on the ground and the hind legs swing in. d): The legs are bunched together and the neck is at its most upright as the right hind leg approaches the ground. (After photos in Dagg and Foster 1982, pp. 100–101)

FIGURE 7.19. A giraffe rising from the lying position.
(Drawings by Jonathan Kingdon 1989, p. 328)

in thrusting forward and contracting into the vertical while landing) are a fascinating sight. The impression that you are watching an animal in slow motion is accentuated during the gallop.

The mechanics of the giraffe's gallop can be understood as follows:

> The power and weight of the giraffe are more in the forequarters than in the hindquarters, so that the main propulsion for each stride comes from the forelegs. By pressing forward at the beginning of each stride, the neck moves into line with the power stroke. The neck facilitates the movement by shifting the center of gravity of the giraffe's body forward and more nearly over the forelegs. At the end of each stride or leg swing, as the hooves touch the ground again, the neck moves backward in order to slow down the forward momentum of the body and enable the giraffe to keep its balance.[57]

In other words, the pendulum motion of the neck helps to propel the giraffe forward and aids in maintaining balance. No other mammal's neck plays such a role in forward movement! And in no other mammal do the forelegs give the main propulsive force, a task usually taken on by the rear legs. Thus the giraffe's unique form of motion arises out of the interplay of its unusual characteristics—its long neck, short body, high center of gravity, and long legs.

The neck not only plays a role in walking and running, but also is absolutely necessary in aiding a giraffe to stand up, as biologist Vaughan Langman describes:

FIGURE 7.20. "Necking" giraffes. (Drawings by Jonathan Kingdon 1989; p. 332)

A giraffe, unlike most other mammals, is totally reliant on its head and neck to rise from lying on its side. In order to get off the ground, it must throw its head and neck toward its legs and use the force of the throw to bring [the giraffe] to its stomach. To come up to a standing position requires another throw of the head and neck, this time back toward the tail; once again it is the momentum of the head-neck throw which makes it possible for a giraffe to stand.[58]

The giraffe's neck, which stands out so conspicuously in a morphological sense, also takes on a prominent role functionally in its movement.

"Necking"

Movement and counter movement appear rhythmical and synchronized, imparting the sinuous grace of a stylized dance.[59]

Imagine a grouping of younger and older male giraffes. One animal starts moving closer to another, until the two are perhaps a few yards apart. He raises his head and neck into an erect posture, emphasizing his height and uprightness. (We might say, anthropomorphically: emphasizing that he's a *real* giraffe.) If the other male responds similarly, they begin walking toward each other, stiff-legged and with legs splayed. They come to stand facing in the same direction, body next to body. They begin leaning and rubbing flanks, necks, and heads against one another. Both giraffes stand with splayed forelegs. One will swing his neck out to the side and swing it back, making contact with the other's neck. The partner responds with the same kind of neck swing. So ensues the "rhythmical and synchronized" dance that Estes characterizes.

This necking behavior, as it is dryly named, can either stop after awhile or transform into a more forceful sparring.[60] In this case the blows with the head and neck become much more powerful, and the slap of contact can be heard far away. When the two giraffes stand side-by-side, but facing in opposite directions, the blows tend to be more violent. Necking bouts may last only a few minutes when one male is clearly dominating the bout. But when the partners are more evenly matched they can last for more than half an hour, and some have even been described as going on for hours. Rarely is a giraffe hurt in these necking bouts; usually one simply stops and wanders off.

Sparring and dominance bouts among males are known in many ungulate species. What is characteristic about this behavior in the giraffe is that the neck plays such a central role. Ecologist Richard Estes speaks of the broad, undulating sweeps of the neck having "sinuous grace." The character of the giraffe comes clearly to expression in this remarkable form of behavior.

Lofty—and at a Distance

Giraffes are not solitary animals. They live in herds of varying sizes, often between 10 and 50 animals. But as Estes puts it,

> The giraffe is not only physically aloof but also socially aloof, forming no lasting bond with its fellows and associating in the most casual way with other individuals whose ranges overlap its own.[61]

Giraffe herds are more accurately described as loose groupings, since their composition continually changes. Groupings rarely stay the same for more than part of a day. In one case, a female giraffe was observed on 800 consecutive days and was only found twice in a group that remained the same for 24 hours. "Giraffe social interactions are highly fluid."[62]

In some cases researchers have observed more long-term relationships. For example, in the Nambian Desert some females show up with other females about one-third to half of the time.[63] That is of course nothing like the bonding between individuals in elephants, but it shows that, perhaps in relation to particular habitats, giraffes groupings are something other than random mixing.[64]

When giraffes are in groups, they tend to keep at a physical distance from each other, remaining within eyesight but often not closer than 20 feet. They reduce these distances when feeding together from the same trees or shrubs. Under these circumstances one can see giraffes closely grouped, although rarely touching each other.

Touching and rubbing are also not typical forms of giraffe social behavior. They occur usually only between cow and calf, between necking males, and before and during mating. Otherwise giraffes prefer distance. You don't see giraffes lounging around with necks resting on the backs of fellow herd members—a typical sight among zebras.

So while there is grouping and some social interaction, giraffes have a kind of self-contained quality. This finds a different expression in their relation to heat. Even on hot days when shade is available, you will often find giraffes in the open. Only when the temperature went over 129°F (54°C) were giraffes observed actively seeking the shade of trees.[65]

It is interesting in this connection that giraffes rarely drink. I have discussed their awkward manner of splaying their forelegs to reach down to drink water, as if their ungainly posture were telling us about their lack of need to drink. They take in substantial amounts of water from the leaves and shoots they browse. Giraffes also do not bathe in watering holes or rivers and rarely swim. If you picture a giraffe immersed in water, it's hard to imagine how it could keep its balance with its high center of gravity. The giraffe is definitely not adapted to life in water!

The quiet, sensitive aloofness of the giraffe stands out more when we contrast it to the elephant. Elephants live in tightly bonded family groups with the members in close physical contact. They rub up against each other and caress and slap each other with their trunks. They are continually pulling in the scents of their surroundings through their trunks. An elephant will smell you before it sees you; its eyes are definitely not its dominant gateway to its surroundings. Elephants also love water and, when they can, bathe every day. Elephants are about contact and immersion; giraffes maintain more distance. Although giraffes and elephants often inhabit the same area, qualitatively they live in very different worlds.

The Developing Giraffe

When people first started reporting about the cow-calf relationship in giraffes, they painted a picture of the all-too-aloof mother. Young calves were often found alone with the mother nowhere in sight for many hours. Over the years, field biologists gained a fuller picture of the cow-calf relationship.[66] A giraffe's gestation period is about 14 to 15 months. Usually one calf is born, dropping to the ground from the considerable height of about two meters. It stands up on its long spindly legs within an hour and then begins suckling. The newborn giraffe is nearly two meters high—just high enough to reach the mother's udder! Since observers find young calves at all times of the year, reproduction appears to be largely independent of the seasons.

Although the calf soon has the ability to stand for longer periods of time and to walk considerable distances, it does not follow its mother while she feeds. The mother leaves the calf in a spot where it is inconspicuous and then wanders off to feed, sometimes going as far as 15 miles. From this behavior the anthropocentric picture of the cold, unconcerned mother arose. But she returns to the calf two to three times during the day and stays with it at night. When she returns, she nudges the calf and licks its neck and the calf begins suckling. When it is done suckling, she leaves again to feed. This pattern continues for about one month.

When the calves are a bit older, different mothers with calves congregate and leave the calves together in crèches during the day. This pattern can go on for six months or more, although the calves already begin supplementing milk with browse after a month. After the crèche phase, the calf stays close to the mother until it is about one and a half years old and then separates for good. A female giraffe becomes pregnant for the first time around four years of age, while males begin mating at about seven.

Calves grow very quickly in the first six months of life, shooting up as much as a yard (about one meter) during this time. After a year the growth rate slows markedly. The adult height is reached after about three to four years.

At birth the legs are—compared to the adult's—proportionally very long and the neck proportionally short. At this time, therefore, the giraffe's neck does not appear particularly long. But during the phase of growth the neck catches up and lengthens considerably faster than the legs. All of the giraffe's other anatomical structures and physiological functions are caught up in this remarkable growth and transformative process.

Feeding Ecology

Giraffes browse primarily on the leaves and twigs of trees and bushes. Rarely do they splay their legs and reach down to feed on forbs; they almost never feed on grasses. Where there is only grassland you don't find giraffes, and desert-dwelling giraffes in Nambia browse along river woodlands.[67] Acacia trees and bushes are one of the giraffe's primary sources of food, so unsurprisingly you are likely to come across giraffes in acacia woodlands and thickets. But giraffes are by no means only narrow acacia specialists (like the koala in Australia, which is bound to eucalyptus trees). Different research groups have found that giraffes feed on 45 to 70 different species of trees and shrubs.[68]

Giraffes are selective feeders; they don't just eat what is right around them. They wander around—up to 20 kilometers per day—feeding on different trees and bushes. They prefer a number of relatively rare species and will browse them more intensively than the much more prevalent acacias that quantitatively make up the largest part of their diet. But even with acacias they are selective, preferring, when they are available, the young shoots and leaves, and in some species the flowers. Because they pick out their food, giraffes spend more time feeding, in relation to their body weight, than, say, their large unselective grazing cousin, the African buffalo. In this respect giraffes are more like the impala, a relatively small antelope.

Giraffes shift their feeding grounds according to the seasons. In East Africa, they are more on the plateaus in acacia savannahs during the wet season, while in the dry season they wander through the valley bottom woodlands that have more fresh browse. Giraffes also have a marked daily rhythm. During daylight they feed or move around, often to find food. Feeding is concentrated in the hours after dawn and before dusk, with a pause during midday. During the day they also ruminate. A giraffe grinds its food using circling motions of the jaw and then swallows it. The movement of the food through the long esophagus is outwardly visible as a bulge racing down the skin of the neck. Soon thereafter the bolus (as the ball of partially digested food is called) shoots back up the esophagus, and you see another wave, this time moving up the neck: a remarkable sight. A giraffe usually stands while ruminating during the day. It has very large salivary glands with which it moistens the cud.[69]

After the sun sets, giraffes feed less frequently. Researcher Robin Pellew found that they feed much more often on moonlit nights than on moonless nights, which suggests that they orient visually when seeking browse.[70] During the dark hours they spend more time ruminating, often while lying down. They also sleep lying down, with their necks curved around and their heads resting on their flanks.

As the nineteenth century British comparative anatomist Richard Owen remarked, "The peculiar length, slenderness and flexibility of the tongue are in exact harmony with the kind of food on which it is destined to subsist."[71] Dagg and Foster describe vividly the browsing giraffe:

> When browsing, a giraffe reaches out with its long dark tongue, wraps the tip about a branch (often heavily thorned), and draws it gently in between its extended lips. Then it closes its mouth and pulls its head away, combing the leaves and small twigs into its mouth with its extra-wide row of lower front teeth. Twigs, leaves, pods, fruit, galls and ants are all chewed together in the tough mouth.[72]

Young shoots have a lighter, fresh green color—a color giraffes probably recognize and use to select just those shoots. The thorns on these shoots are still green and flexible, but giraffes have no problem feeding on more mature shoots with sizable thorns. Autopsies of stomachs of wild-killed giraffe reveal leaves, twigs, seeds, and thorns up to three and a half centimeters in length.[73] The tongue's upper surface is thick-skinned and covered with small spines, giving it a sandpaper consistency. The inside of the mouth is also clothed with a tough epidermis. So the giraffe's discriminating style of feeding, with the dexterous motion of the agile tongue, is infused with an underlying toughness.

Giraffes can feed on longer-thorned acacia species "with remarkable skill and care, using the extremely mobile tongue in conjunction with the soft lips."[74] In the process they not only tear off leaves but also break off

FIGURE 7.21. A lone male giraffe in Botswana. Note that the tail is missing its long hairs, which were probably lost when it was grabbed by a lion.

thorns, which pass through the well-protected mouth, down the esophagus, and into the stomach, where they are at least in part digested. Nonetheless giraffes prefer younger, soft-thorned shoots. In a field experiment scientists removed thorns from branches of acacia trees (*Acacia seyal*) and found that giraffes browsed these dethorned branches significantly more intensively than they did the naturally thorned branches.[75]

Although giraffes are clearly predisposed to selective feeding, they do survive quite well in zoos where they are fed much more fiber-rich, low protein diets. (Young leaves have a higher protein content than old leaves.) Hofmann and Matern[76] performed autopsies on zoo and wild giraffes and found remarkable differences in the first two, largest, chambers of the stomach (the rumen and the reticulum). The zoo giraffes had significantly larger chambers, which could hold about 150 liters (nearly 40 gallons) as compared to an average of 105 liters (nearly 28 gallons) in wild giraffes. This change correlates with the increased amount of roughage in the feed of zoo giraffes. Grass-eating (grazing) ruminants, which ingest more roughage, have proportionately larger stomachs than broad-leaf-eating browsers. So it is not too surprising that when giraffes are fed more like grazers, their stomachs enlarge to accommodate the digestion of high-roughage feed.

But with the increase in stomach volume in the zoo giraffes, the absorptive surface also decreased, which was shown through the small size and number of the papillae that make up the inner lining of the rumen. The wild giraffes, in contrast, had much larger and more numerous papillae, so that their absorptive surface was in fact nine times larger. This allows for the intensive digestion of the fresh material they feed on, which passes much more quickly through the digestive system than roughage-rich fodder.

This remarkable adaptation of the stomach to the food a giraffe eats shows that the giraffe is not physiologically or anatomically set in its ways. It is flexible and can adapt to changing conditions.

The Intertwined Existence of Acacia and Giraffe

In their classical umbrella form—a broad, spreading crown branching off from a single erect trunk—acacias help define the savannah landscape in which giraffes thrive. We have seen how giraffes live off acacias as a primary source of food, but the interaction between these two very different organisms goes deeper.

Acacias in Australia—which is home to nearly one thousand species—are thornless, whereas virtually all of the approximately 130 species in Africa bear thorns. Since there are no large herbivores in Australia that browse acacias, it's obvious to think that African acacias might have developed thorns in response to the browsing of giraffes and other browsers. Field biologists have made observations that support this view, noting that acacia branches above the height of giraffe browsing have fewer and shorter thorns than the branches accessible to giraffes.[77] When giraffes and elephants were excluded from areas of acacia woods in field experiments, the new shoots developed *shorter* thorns.[78] Browsing may not explain the origin of thorns in African acacias, but it is certainly evident that the length and extent of thorns is influenced by browsing, with giraffes playing a primary role.

It is a simple matter to picture thorn formation as an adaptive response that keeps browsers from feeding on a tree, thereby increasing its survival chances. The interesting thing is that giraffes feed on acacias even when they are densely packed with thorns. The coevolution of thorn formation and giraffe browsing does not lead these two organisms to sever their interaction. Maybe thinking of thorns solely as weapons to deter browsing is too narrow a view. Although we don't yet know it, there may be more to thorns than pricks a mouth.

Similarly ambiguous is the evolution of stinging ants that live exclusively on the whistling thorn (*Acacia drepanolobium*) in East Africa. These ants build their nests in the bulbous swellings at the base of the modified thorns of this tree. Only this species of acacia has these swellings that can provide such a home for ants. The ants live from the nectar produced in the acacia's leaves. When an antelope or a giraffe browses on a branch, the ants swarm out and sting it. Again, thinking simplistically of "stinging ants as acacia protectors," you might say that the stings ward off the browsers. In fact, one study showed that trees with more active ants had more foliage than those with less active colonies. Moreover, young giraffes browsed less on trees with more active ants. But the ants had no apparent effect on the browsing behavior of adult giraffes.[79] Once again, reality is more complex than simple (and convenient) explanations.

When we view thorns and ants exclusively as defensive mechanisms, we assume the acacia and the giraffe are antagonists, each busily shaping the survival of its own species. We view species as separate entities that interact on the basis of competition. But species are not separate entities; every species lives from and provides life to many other kinds of organisms. When we view species interaction in terms of coexistence, where each species, through its own life, supports the life of other species, we transcend the narrow terms of competition and individual species survival that constrain so much of ecological and evolutionary thought today.

Scientists in South Africa observed that giraffes browsed acacias near water holes more intensely than trees far away from such water sources.[80] Acacias grow new shoots after the onset of the rainy season (one or two times a year). The scientists observed that the shoots from the more heavily browsed trees grew back very rapidly, and grew longer, which compensated for the intense browsing. In contrast, the lightly browsed acacias grew smaller shoots, so that the net shoot extension was the same in both habitats. In other words, giraffe browsing stimulated growth of the acacias in relation to the degree of browsing—a wonderful example of dynamic balance (which is disturbed when the habitat is too small for the number of giraffes living in it). This interaction is an example of a widespread phenomenon in plants known as compensatory growth.[81]

The heavily browsed acacias reacted to giraffe feeding in another, perhaps more surprising way. The leaves that grew in the rainy season after browsing were richer in nutrients and contained significantly less condensed tannins, which make leaves less palatable. Tannins are formed after cessation of leaf growth, while nutrient-rich phosphorus and nitrogen compounds are formed during growth. Stimulated by browsing, the acacia leaves remained in a more juvenile state, which is exactly the type of leaf giraffes prefer!

Of course, a population of herbivores can become too large for a habitat and damage it. Usually this happens when human beings limit the movement and therefore the home range of the animals. For example, in Ithala Game Reserve in South Africa, five giraffes were introduced in 1977, and their population increased to around 200 by 1999. Giraffes browse intensively in a relatively small area (they cannot navigate the steep slopes in the preserve); here one species of acacia has disappeared altogether.[82]

Let me mention one more example of the intertwined biology of giraffes and acacias. The flowers of knob thorn (*Acacia nigrescens*) are an important source of food for giraffes in the late dry season in southern Africa, when many trees have lost their leaves.[83] The flowers of this species grow bunched on stalks (they have a "bottle brush" form) and stand beyond the fairly small, curved thorns. Most of the flowers consist only of pollen-producing stamens and lack the pistil out of which fruits and seeds form. Giraffe browsing has little to no effect on the reproductive capacity of the tree—what the tree offers in abundance the giraffe takes.

An Integrated Being

It's not by chance that all people who study the giraffe are brought back again and again to its long neck. What we have seen is that the long neck is the most vivid and perhaps most extreme manifestation of an overall tendency toward vertical elongation in the giraffe. We see this feature not only in the neck but also in its long slender legs, in its long head, its long tongue, and even its long heart. The vertical orientation of the body is increased by the shortness of the rump, which slopes upward into the neck. Vertical elongation reveals itself most prominently in the front part of the animal; the giraffe is the only hoofed mammal with longer forelegs than hind legs. The forelegs extend into the long and sleek shoulder blade. Because the neck also emerges high up on the trunk, its upward reach is extended even more. This augmented elongation makes it impossible for the giraffe to reach the ground with its head without spreading or bending its forelegs. Truly, the giraffe has a one-sided, but highly integrated morphology.

With its long strides the giraffe ambles through the savannah. The neck helps propel the animal forward when it transitions into a gallop. Similarly, when standing up, the giraffe throws its neck upward to help lift its body from the ground. And in male neck sparring, the sinuous beauty and power of the neck shows in yet another way the dominance of this organ in the life of the giraffe.

E.-M. Kranich points out that the neck is the organ in an animal that frees the head from its close connection to the unconscious vital processes of the rump. Thereby the head becomes a more autonomous center of sensation and perception.[84] With its long neck, the giraffe carries its head significantly higher above the rump than does any other mammal, opening itself to the wide surroundings through its large eyes and keen vision. It is a calm, silent sentinel. With its overviewing eyes, it lives more like a bird surveying its surroundings from on high than like so many of its mammalian relatives that bathe, head-lowered, in the near-ground world of scents. Characteristically, the giraffe has a dry snout and not the moist snout of a typical mammal, upon which airborne scents can dissolve. Giraffes communicate with one another largely through vision, much less through touch and smell.

Like other browsers, the giraffe spends much of the day feeding, selectively browsing on an array of trees and bushes, but with a decided preference for acacias. Its elongated neck, head, and tongue give the giraffe a uniquely large vertical span within which it can browse (around 13 feet). It can also reach far in front of its body using the neck as an immense arm to gather food at a distance. And, of course, at times the giraffe reaches with neck, head, and tongue in line to great heights to feed. With its flexible and yet tough tongue, it reaches out to enwrap, if possible, young and tender leaves. But it is also impervious to woody, pointed thorns. The ingested acacia becomes part of the giraffe, and giraffe browsing affects, in turn, the growth of the acacia. Through such interrelationships we see that the giraffe as an organism is part of the larger organism of the environment.

The Idea of Evolution

The idea that organisms evolve began to take hold of human minds in the decades before and after 1800. This spurt in interest was not because a wealth of new evidence for evolution was suddenly laid out. Rather, individual thinkers and scientists began viewing geological, biological, and historical processes in terms of development and transformation.[85] What had previously been looked upon in biology as separate entities— species, genera, etc.—related ideally in a "great chain of being," became in the minds of early evolutionary thinkers members of an unbroken stream of evolutionary transformation. For example, Johann Gottfried Herder wrote in 1784:

Compounds of water, air, and light must have arisen before the seed of the first plant-organization, like the mosses, could arise. Many plants must have arisen and then died before an animal-organization became. Also insects, birds, water and nocturnal animals must have preceded the advanced animals of the land and day.[86]

The idea of evolution shed light on things; it was revelatory. And what more does the human mind seek in the search for knowledge than ideas that illuminate the nature of the world we are investigating? This revolution in human thought was not bound to any particular theory. Whether promoted by materialists or individuals who believed in a spiritual foundation to the world (like Herder), by scientists or philosophers, the idea of evolution caught on. The intuition that things evolve was the wellspring for new ways of viewing the natural world and human history. Specific interpretations and explanations of evolutionary processes came second. Differing views of evolution arose, depending on the perspectives of the individuals. Some were spiritual, others materialistic; some were teleological, others emphasized randomness; some placed structure in the foreground, others function.

After the publication of *Origin of Species* in 1859, Darwin's cogent and well-argued theory of evolution became, over time, *the* theory of evolution. Most of us today, when we hear the term "evolution," think immediately of the Darwinian theory of evolution: random mutation and natural selection drive the evolution of species. We probably don't even know that there always have been and still are other ways of interpreting evolutionary phenomena. There has been a resurgence of alternative perspectives in the recent past.[87]

In the United States we seem to have a propensity toward dichotomies, so that someone is either a "real" evolutionist (meaning a Darwinist) or a Creationist (i.e., someone who doesn't believe in evolution). Some scientists and philosophers have formulated what they call "intelligent design" as one attempt to wed spiritual and evolutionary views;[88] they are, by and large, pushed into the Creationist camp by Darwinians. The battle between Darwinists and Creationists, fought on both sides with religious fervor, has led to unfortunate oversimplifications and to an unwarranted polarization of perspectives. This dichotomizing is, to my mind, counterproductive and shuts down our thinking about some of the true riddles of evolution.

So what can we do? We can step back and look more openly at the phenomena themselves and the challenges they present to us. What emerges is a many-faceted picture of evolution that leads to new kinds of questions. In my own studies of evolution I have found so many doors closed through narrow ways of viewing that I have come to see the most important step entails opening up some of those portals and letting in fresh air. This approach may not provide the safe surroundings of a closed, coherent system, but it is invigorating.

Okapi and Giraffe

The skin of an unknown, horse-sized mammal from the central African rainforest (the Ituri Forest of Congo-Zaire) was sent to Europe in 1900. The skin was dark brown, almost black in areas, but had zebra-like stripes on the legs and rear quarters. Was it a rainforest zebra? From the skin alone there was no telling. Since it had been acquired from the Pygmies living in the forest, they were asked about the animal. They insisted that it has paired hooves and not a single hoof like a zebra's. They also described its large donkey-like ears and the small, spiked-shaped and hair-covered horns the male carries. So it was definitely not a zebra.

Soon a skull of the animal arrived. The horns were like those of a giraffe, and the skull had two-lobed canines in the lower jaw, which only giraffes possess. So this newly discovered mammal was a member of the giraffe family! The new species was called okapi, after the Pygmies' name for it, and then given the scientific name *Okapia johnstoni*.

Based only on the skull and skin, artist P. J. Smit, under the guidance of anatomist and British Museum of Natural History Director Sir Ray Lankester, painted a reconstruction of the okapi that proved to be an almost exact likeness to the living animal. This is a remarkable example of how careful examination of a limited number of parts coupled with a broad and deep knowledge of the anatomy and morphology of comparable animals can lead to a picture of the whole animal.

This "extraordinarily handsome animal"[89] was seen alive by Europeans only twice before 1906. For most of the twentieth century, scientists could observe the okapi only in zoos, until in the last few decades field

FIGURE 7.22. An okapi (*Okapia johnstoni*). (Drawing by Jonathan Kingdon 1989, p. 338)

researchers began to learn more about this shy and elusive animal's life history and ecology.[90]

The okapi has aroused additional interest because its skeletal structure is very similar to that of fossils discovered in Asia and Africa. It quickly became known as a "living fossil," although that label is controversial among specialists. The question arose, Does the okapi give us a glimpse of the ancestor of giraffes?

Clearly, the okapi does not have the giraffe's characteristic long neck and short body. Both its neck and body are more like those of an antelope or deer than the giraffe's. In its overall form, the body is less specialized than the giraffe's. But it does have long legs. While in other ungulates the rear legs are significantly longer than the front legs, in the okapi they are nearly the same length. This is a step in the direction of the giraffe. Moreover, the okapi, like the giraffe, splays its forelegs when grazing near the ground. As another feature of elongation, the okapi also has a long, pointed, and flexible tongue. So in some respects we can see in the okapi the nascent giraffe. But to know more, we must turn to the fossil record.

Fossil Giraffes

The fossil record is a picture in the present of life in the past. We find traces of life in fossilized bones, imprints, and other fossilized body parts and can build up pictures of animals, plants, and habitats of the past. Of course these pictures are always subject to alteration, and at times our fantasy will take flight. The skins of animals, for example, are almost never found, yet reconstructions usually present animals in full color and patterning. We need to be careful that our pictures remain tentative and open. With this in mind, let's look at the fossil history of the giraffe family, known to scientists as the giraffids (Giraffidae).[91]

If the evolution of the giraffe had progressed as Darwin envisioned, one would expect to find fossils of many intermediate stages of animals with successively longer necks and legs between the giraffe ancestor—a small deer- or antelope-like animal that perhaps resembled the okapi—and the fully evolved giraffe. But this is not the case.

Fossils of giraffes—perhaps not the same species as today's *Giraffa camelopardalis*, but clearly giraffes—can be found in Europe (Greece) and Asia in the layers of the lower Pliocene and the upper Miocene, geological periods that directly precede the last ice age (the Pleistocene period). In these strata, one finds fully developed giraffes—some smaller, some larger and more robust—along with other, now extinct, members of the giraffe family. That these other fossils belong to the giraffe family and not, say, to the deer or cattle families, can be seen in such diagnostic characters as the bilobed lower canine teeth and the horns.

Up until now, fossils of giraffes in Africa have been found only in more recent layers. These earliest known remains of a giraffe (called *Bohlinia* or *Orasius*) closely resemble modern giraffes, both in size and shape. Key features of the skull are very similar to the present-day giraffe, and limb length and proportions "agree fully with *Giraffa*," writes paleontologist Birger Bohlin, who described the fossil remains in detail.[92]

Two groups (subfamilies) of giraffids coexisted with the early giraffes in Africa, the Palaeotraginae and the Sivatheriinae. The former were generally deer- to elk-sized animals with long legs and body proportions resembling the okapi. The sivatheres consisted of massive, in some cases, elephant-sized animals that were much stockier than other giraffids. *Sivatherium maurusium* resembled an elephant-sized moose (see Figure

FIGURE 7.23. Representatives of the three members of the giraffe family: a reconstruction of the extinct Sivatherium, the okapi, and the giraffe. The drawings are to scale. (Drawings after Churcher 1978 [Sivatherium], Kingdon 1989 [okapi], and Skinner and Smithers 1990 [giraffe])

7.23). It even had spreading horns that resemble antlers. In both subfamilies there was a wide array of horn forms. Neither the palaeotragrines nor the sivatheres had the unusual limb proportions, the short torso or the overly elongated neck of the giraffe.

So there were three quite distinct groups of giraffids: the large, long-legged, short-bodied, and long-necked giraffes; the massive sivatheres; and the more "typical" ungulate group to which today's okapi belongs.

Before giraffes appeared, one finds many fossils, both in Africa and Asia, belonging to the okapi-like Palaeotraginae subfamily. There is diversity among these fossils, which has led paleontologists to speak of three different genera (*Giraffokeryx*, *Palaeotragus*, and *Samotherium*). All have quite long limbs, and the forelimbs had become about the same length of the hind limbs. The proportions of the lengths of individual bones in the limbs resemble those of the okapi much more than those of the giraffe. However, some species (e.g., *Palaeotragus germaini* and *Samotherium major*) have somewhat elongated neck vertebrae that resemble in morphology giraffe vertebrae.[93] The paleontologist Arambourg considered this tendency toward neck lengthening within the Palaeotraginae as a parallel evolution to what was developing, perhaps at the same time, in the—still unknown—predecessors of the giraffe.[94]

The remains of the first giraffids are found in the lower layers of the early Miocene fossil record of Africa. One species, *Canthumeryx sirtensis* (formerly called *Zarafa zelteni*), resembles a "lightly built deer or antelope, with generally slender proportions and light build to all parts."[95] It is generally viewed as the most "primitive" giraffid, since it is both an early representative and its body is not highly specialized. It comes the nearest to being the basal species of giraffids. Because, however, it coexisted with a relatively small sivathere, it is not considered to actually be at the base of giraffe family evolution. That origin, as is so often the case when one arrives near the base of an evolutionary tree or bush of a given group of animals, remains dark. Interestingly, the giraffe family evolved later than other ungulate families.

What the fossil record does not show is intermediate forms linking early okapi-like animals with the giraffe and its specialized morphology. The fossils tell no clear-cut story. Three quite distinct subfamilies evolved with considerable variation within these groups. But if you're looking for the gradual transition of one form into another, a picture suggested by Darwin's view of evolution via the accumulation of small variations over long periods of time, the fossil record is disappointing. One might argue that this is an artifact of the incompleteness of the fossil record. And, of course, no one can predict what still might be found.

A Temporal Pattern of Development

Increasingly paleontologists recognize that the lack of intermediate stages between related groups is a *typical pattern* within the fossil record. In many groups of animals, the fossil record is characterized by the development of various distinct subgroups that coexisted over long periods of time. In other words, a new group (family or genus) evolves rapidly and then exists for a longer period of time characterized by small evolutionary modifications. The German paleontologist Otto Schindewolf, one of the first to recognize this gestalt of the fossil record, spoke of two evolutionary phases: typogenesis, in which the new group appears, and then the much longer period of typostasis, when the group evolves in small increments without radically new characteristics developing.[96]

In 1972, American paleontologists Niles Eldredge and Stephen Jay Gould formulated basically the same idea, describing evolution as a process of "punctuated equilibrium" in which long periods of relative morphological stability are punctuated by evolutionary innovations.[97] As one eminent paleontologist, Robert Carroll, writes: "Instead of new families, orders, and classes evolving from one another over long periods of time, most attained their distinctive characteristics when they first appeared in the fossil record and have retained this basic pattern for the remainder of their duration."[98]

This picture certainly fits the current fossil-based evidence of evolution within the giraffe family better than a gradualist one. It seems likely that the main thrust of giraffe evolution occurred in a condensed period of time, followed by a longer period, extending to the present day, with relatively little change.

If we look around us at developmental processes today we also find that major changes usually occur rapidly. Embryonic development is the example par excellence. The most significant events in biological development occur in a short period of time within the overall span of the organism's life. In the first nine weeks of human embryonic development, for example, a tiny one-celled fertilized egg becomes a small (about one inch-long) embryo that has already begun to develop all of its organs. These nascent organs then differentiate further until birth and beyond. Never again does our body go through such rapid, all-encompassing transformation—nine months in a lifetime of perhaps sixty to eighty years.

Postnatal development is also characterized by periods of greater stasis and ones of more rapid change—growth spurts, puberty, and menopause, to name a few. Even in a period of rapid change, such as the fast growth of infants, growth is not evenly distributed but occurs in bursts. For example, the growth of infants ranged between a quarter of an inch and one inch (0.5 and 2.5 centimeters) over a 60-day period, and researchers found that most of this growth occurred during single nights—up to two-thirds of an inch (1.65 centimeters) in a night![99]

There are other phenomena of development that we can observe intimately, but tend to overlook. I mean the development of knowledge. Just think of when we have an "Aha!" experience, a new insight that sheds bright new light on things. Such new insights are of course usually borne out of strenuous efforts—perhaps longer periods of time in which nothing "comes"—and then at once a new insight is there. It does not simply grow gradually in incremental steps by adding on to past knowledge; it is a new idea that reorganizes our past knowledge, revealing new relationships and connections—and giving rise to new questions. Our body of knowledge takes on a new gestalt.

In these examples we see both the temporal dynamics of a developmental process and the fact that when rapid changes occur they affect the whole system and not just some isolated part. The metamorphosis of a tadpole into a frog, which I will explore in detail in the next chapter, is a wonderful example of such an all-encompassing organic transformation. The fish-like, fin- and tail-bearing, gill-breathing, herbivorous tadpole totally transforms in a short period of time into a strong-legged, tailless, lung-breathing carnivore. Every organ goes through dramatic changes as the frog develops by the reorganization of everything that was previously "tadpole."

This points us to another feature of such transformations: they leave no remnants. The whole system reorganizes. This makes sense, since we are dealing with an organism in which all parts (as members) are interconnected. Major developmental steps are not about incremental additions to a preexisting stable structure that remains essentially unchanged.

So if an organism evolves as an organism, and not as a collection of parts, then the pattern in the fossil record indicating that major evolutionary steps occur rapidly is actually not so surprising. It is a time gestalt or pattern on a large scale that we find everywhere within developmental processes we can directly perceive today—dynamics involving phases of accelerated change and phases of greater stability. From this perspective of major steps in evolution (often called macroevolution) as a developmental process, the "gaps" in the fossil record appear as a consequence of a rapid reorganization that leaves no remnants, no physical tracks.

I am certainly not suggesting that one day an okapi-like animal gave birth to a fully developed giraffe. How a new animal group in the past arose remains a huge riddle. Here we come up against a boundary of our present-day understanding. I prefer to acknowledge that boundary and not to begin theorizing and speculating about how, in concrete terms, this process might have taken place. All possible "mechanisms" that purportedly explain macroevolution end up replacing the true complexity of the matter with a simplified scheme. It is more fruitful, I believe, to be conscious of the boundary, hold back on speculative explanations, and continue to explore the rich contexts within which evolution occurs.

An Overriding Morphological Pattern

We have just seen how the fossil record of the giraffe family indicates a time pattern within more general developmental processes. While researching the giraffid fossil record, I was struck by another type of overriding pattern that German biologist Wolfgang Schad has found in living mammals and described in great detail.[100] Schad found that many groups of mammals fall quite naturally into three subgroups. For example, three ungulate families (the so-called pecorans) have horns or antlers: the cattle, deer, and giraffe families. Similarly, the odd-toed ungulates also fall into three families: the rhinoceroses, tapirs, and horses (which include zebras). What's interesting is not so much the numerical pattern but that within each of the groupings you find a biological polarity mediated by an intermediate form.

I already pointed to this pattern in the comparison of bison, elk, and giraffe. The bison is a member of the bovid family (which includes the cow, yak, sheep, goat, and antelope). On the whole these are weighty, compact, and bulky animals. Think of the bison with its short legs and heavy, low-to-the-ground head. In contrast, the giraffe is tall and sleek with long neck and legs. The deer family, represented by the elk, has a more balanced, intermediate form. In a related way, within the odd-toed ungulates, the rather unspecialized tapirs represent a mediating form between the massive rhinoceros and the relatively long-necked and long-legged, swift-footed horses and zebras.

What's intriguing is that this pattern of morphological extremes with an intermediate group is reiterated within each group. So within the bovids you have the cattle group (bison, cow, yak) on one side and the fleet-footed, sleek antelopes on the other, with sheep and goats in between. Similarly in the deer family, there are the more petite deer species such as the European roe deer or the somewhat more robust American white-tailed deer on the one hand and the large, bulky moose on the other hand: the elk (wapiti) represents a middle form within this group.

After I had studied the giraffe family fossil record for some time, it came to me that this pattern was showing itself again within this group. There are, as we have seen, three groups of giraffids. The sivatheres, which are extinct, were large, sometimes huge, animals. Proportionately they have the shortest legs and neck of all giraffids. But they often have massive antlers resembling those of the moose, which we just saw also falls to one pole of its family. The long-legged giraffe represents the other pole with its long neck separating its head from its shortened body. The okapi-like fossils form an intermediate group with less extreme specializations.

It would lead us too far afield to go into a large number of examples here to show all the variations, iterations, and nuances of this threefold pattern as they show themselves in form, coat patterns, physiology, and behavior. Since Schad's groundbreaking study, others have found a similar pattern not only in mammals, but also in dinosaurs, birds, insects, and other groups of animals.[101] Of course this is not the only kind of morphological pattern to be discovered, and you have to be cautious that you don't end up fitting everything into a neat scheme that no longer does justice to the phenomena. We need to hold such patterns in as flexible and open a form as possible.

But acknowledging such overriding morphological patterns has important consequences for how we think about evolution. Finding patterns that encompass many groups of animals indicates that we need to go beyond a focus upon a given species or genus to understand evolution. The threefold pattern suggests, at a macroevolutionary level, the evolution of a kind of "superorganism" that differentiates into extremes (poles) and a middle form within a given class, family, or other systematic group. This notion is, for our standard ways of thinking about evolution, quite radical. But it is one suggested by the order of the phenomena themselves. The problem is that we have to stretch our thinking beyond the idea of wholly self-contained organisms and begin to see each species, genus, or family as embedded within a larger organismic context that encompasses many animals.

Ecology and Evolution

When we look at the ecological relationships between organisms today, it is clear that the life of species is intimately intertwined. Because it is impossible to understand any organism in isolation, ecologists have found it necessary to take concepts that were originally conceived of in connection with individual organisms and expand them to refer to larger ecological categories. They use terms such as "ecosystem genetics" and "community heritability" to express how a whole ecosystem evolves with the individual species being members or organs within the larger system.[102]

The giraffe evolved within the context of a savannah environment.[103] We have seen how in present times the giraffe is intimately interwoven with its savannah habitat in Africa. It lives, for example, from the acacia, which is modified by the giraffe and in turn affects the giraffe. We can only envision evolution as coevolution. No creature evolves by itself as if in a vacuum. Just as we can see that the giraffe evolves as a member of an organic threefold pattern with respect to other animals, so also we can see that its evolution is nested within an environment, a larger organism, that in the end encompasses the whole earth. Goethe, who was an eminently ecological thinker, expressed this view already in the 1790s:

> We will see the entire plant world, for example, as a vast sea which is as necessary to the existence of individual insects as the oceans and rivers are to the existence of individual fish, and we will observe that an enormous number of living creatures are born and nourished in this ocean of plants. Ultimately we will see the whole world of animals as a great element in which one species is created, or at least sustained, by and through another.[104]

For a great variety of mammal—think of lions, hippos, giraffes, impalas, or elephants—the African savannah is their womb and sustenance. They all belong together in this landscape.

Savannah-like conditions exist in other parts of the world, but only Africa sustains such a diversity of large mammals. What was and is at work specifically on this continent? This question of the specific qualities of the different continents and bioregions looms large. It goes much deeper than the sum of climatic and geological factors. Similarly, the question of the unique quality of each animal encompasses more than the sum of genetic and environmental factors.

Nested Contexts

In the previous sections I have tried to show that we can begin to understand the evolution of the giraffe only if we view it within larger contexts. One context is the fossil record of the giraffe family. It points to a temporal pattern that initially is surprising: there is no series of connecting links between the different subfamilies, showing how the giraffe might have gradually evolved in a step by step fashion. Rather, the three subfamilies appear as quite distinct groups. The major evolutionary innovations apparently occurred quite rapidly (macroevolution), while over longer periods of time smaller variations within a given group arose (microevolution). This pattern appears to be widespread in the fossil record of other organisms and is also typical of the way developmental processes occur today. In this sense the giraffe's evolution is part of a more general evolutionary or developmental trend.

A second context is a morphological one: the threefold pattern of differentiating into two extremes and an intermediate group. We find this pattern not only in the giraffe family but also in many other groups of animals, both vertebrates and invertebrates. It is an overriding morphological pattern of which giraffe evolution is a part.

A third context is ecological. All life is interconnected, and the evolution of any organism is bound up in the evolution of others. Evolution at this level entails the modification of already existing forms and relates mostly to microevolutionary processes.

The fourth context—and I know it may sound strange putting it like this—is the giraffe itself. It has been the main focus of this chapter. We can view the giraffe as evolving within the context of more encompassing patterns—as we have seen in the threefold differentiation that occurs within myriad systematic groups. The giraffe family, within the order of the even-toed ungulates, shows a distinct tendency toward limb lengthening. But the tendency toward vertical lengthening that plays itself out to an extreme in the giraffe, with the whole body form and structure being reorganized in relation to the disproportionate lengthening of the neck, cannot be deduced from the overriding pattern.

No one could predict that the giraffe would evolve in just the way that it has.

Similarly, we cannot derive the giraffe's unique qualities from environmental interactions. As Mitchell and Skinner point out, the giraffe's "physiological adaptations ... subserve the needs imposed by their anatomy rather than the needs imposed by their environment."[105] They view the giraffe as a unique case because of its extreme specializations. But actually they are directing us to a reality that holds for all organisms, namely the primacy of the coherence of the organism itself. You have to think about the organism in its own terms and then see the manifold relations to the environment.

Back to the Whole Organism

If we do not have at least an educated inkling of what an organism truly is, it becomes all too easy to explain the organism away through neat evolutionary stories. With these stories we "educate" our high school and college students, and the public via the popular press.

The evolutionary stories that I discussed at the beginning of this chapter and that want to explain how the giraffe evolved its long neck, now appear almost laughable in light of a richer, contextualized knowledge of giraffe biology, ecology, and the fossil record. The inevitable shortcomings of such evolutionary stories have to do with focusing on particular characteristics (long neck) and considering them in isolation from the rest of the organism. You then take a next step and pick out a particular function (selecting it from the various functions that any part of an animal has) and consider it from only one narrow perspective: how it contributes to the animal's survival. The trait (long neck) becomes a survival strategy (allows survival during droughts), and on this basis you build your picture of evolution.

Both the trait ("long neck") and the particular survival strategy are products of a process of abstraction from a complex whole. Therefore we can think them clearly and establish a clean and transparent explanation that seems to work—it makes sense that the long neck evolved in relation to survival and feeding habits. Unfortunately—and it is the price we pay when we operate with abstractions—we have lost the giraffe as the whole, integrated creature it is. We're not explaining the giraffe; we're explaining a surrogate that we have constructed in our minds. The animal has become a bloodless scheme. For this reason evolutionary stories are usually woefully inadequate. This way of viewing evolution and the organism is simply too narrow, too one-sided, and too unaware of its own limitations.

Anthropologist and historian of science Loren Eisely points to the central perspective that is largely missing in the Darwinian approach:

> Darwin's primary interest [was] the modification of living forms under the selective influence of the environment.... Magnificent as his grasp of this aspect of biology is, it is counterbalanced by a curious lack of interest in the nature of the organism itself.... It is difficult to find in Darwin any really deep recognition of the life of the organism as a functioning whole which must be coordinated interiorly before it can function exteriorly. He was, as we have said, a separatist, a student of parts and their changes. He looked upon the organism as a cloud form altering under the winds of chance and it was the permutations and transmutations of its substance that interested him. The inner nature of the cloud, its stability as a cloud, even as it was drawn out, flattened, or compressed by the forces of time and circumstance, moved him but little.[106]

In a sense, Darwin's gift of seeing how organisms change in relation to their environment kept him from acknowledging the organism in its own right. In and of itself this is not a problem. The problem arises when a one-sided approach becomes the only approach. Since Darwin, the main body of evolutionary science has followed this narrow and one-sided pathway. With the focus on genes and traits as survival strategies, the organism itself has been virtually lost from view.[107]

Recognizing this problem, I have attempted in this study to give a many-sided picture of the giraffe. My intent here has not been to "explain" in any narrow sense (which would entail reducing complex phenomena to underlying mechanisms). Rather, I have tried to present a comprehensive picture and let it stand as such. Such a picture brings us closer to the giraffe and its uniqueness, revealing the broader contexts of which the giraffe is a member, and also showing us what we don't yet know.

My central aim has been to bring the giraffe as a whole organism into view—to show the interconnectedness of its features and how we can begin to grasp it as an integrated whole. When we take the organism seriously, we gain knowledge of its integrity. This is what I have attempted with respect to the giraffe and its long neck. Yes, a giraffe can reach great heights to feed. This ability is remarkable, but it does not explain the neck. When we study the giraffe more in detail and the neck within the context of the whole animal, we discover an overriding tendency toward vertical lengthening that comes to fullest expression in the neck. This tendency shows itself in the legs, in the head, and in the shortening of the body. It becomes a key to understanding much that is unusual and special about the giraffe: the way it stands up, walks, and runs; the way it can reach so high; the way it awkwardly spreads its legs to drink; the way males spar with their rhythmically batting necks; its sensory focus in the overviewing eye. All these characteristics reveal the inner coherence of the organism. The picture of a specialized creature emerges with its unique—and one-sided—characteristics.

Even a nascent understanding of this holistic character of the giraffe transforms our image of what it means to be an animal. The giraffe is not a composite; it is not put together puzzle-like out of separate components. Its "long neck" is not a discrete trait added onto an already existing edifice.

From this perspective, the animal is an end in itself and irreducible. Typical evolutionary "explanations" ignore this quality of the organism and go to great lengths to reduce all evolutionary processes to genetics and environment. By so doing they forget the organism as the context of action and reaction for both genes and environmental cues. Unless you presuppose a center of cohesive activity that is the evolving organism itself, you are dealing with an abstraction.

The pillars of contemporary Darwinism—genetic mutation, gene recombination, and natural selection—therefore appear as modifying, regulatory factors and not as the driving forces of evolutionary innovation. They presuppose the "organism as a functioning whole ... coordinated interiorly before it can function exteriorly," to use Eisely's expression. The whole organism—conceived in broadest terms—is the context for both genes and natural selection and is not a mere effect of their actions. It is the crucial font of evolutionary innovation.

Placing the holistic gestalt of the organism back into the center of evolutionary considerations has enormous implications. We can no longer look at animal evolution as the outcome of the interaction of causal mechanisms. We are always led back to beings that evolve in relation to other beings and to the many inorganic forces of nature. Beings interact and coevolve; yet each evolves in its particular way. No being is reducible to something else. This is a central riddle of life on earth, of development, and in the end, of evolution.

* * *

A towering giraffe ambles over a grassland, its long tail swishing back and forth while its deep black eyes glisten alertly. Its gaze encompasses a broad horizon high above the sea of ground-level smells in which other mammals live. The short body carried high by its long slender legs surges up into the grand neck from which the giraffe looks mostly down upon the world. Moving into the trees and bushes its clear gestalt dissolves among the patches of dark and light. It lowers its neck, extends its long tongue and enwraps a branch, stripping off the young leaves. It prunes the bushes and small trees from above and, occasionally, stretches its neck, head, and tongue upward to reach the lower branches of a large acacia tree, pruning it from below.

When we get too close the giraffe moves off and breaks into a gallop. It spreads its long legs and brings them back together in long strides. With its head held high, the neck undulates slightly to and fro. This ethereal movement stops and the giraffe stands. Its gaze spreads into the world around it.

CHAPTER 8

Do Frogs Come From Tadpoles?

WHERE DOES A FROG COME FROM? The answer seems obvious. It comes from a tadpole. But does it?

Surely, without the tadpole the frog does not develop. But just as surely, nowhere do we find the frog in the tadpole. You would never guess, if you were familiar only with tadpoles, that they would turn into frogs. And the frog comes into existence only as the tadpole disappears. We need to be keenly aware of what we mean, and what we don't mean, when we say, "A frog develops out of a tadpole"—or a tadpole out of an embryo, or an embryo out of a fertilized egg, or an adult human being out of a child.

As we will see, when we give careful attention to what is actually happening when a new phase of life develops out of a previous stage, there are large implications for our overall understanding of developmental processes. New and exciting questions arise about how we conceive of development—including that trans-species developmental process we call evolution. But first we need to get to know frogs a bit.

FIGURE 8.1. A green frog (*Rana clamitans*).

At Home in the In Between

If you live in a temperate climate, you have most likely heard frogs croaking at the edges of ponds, or seen them swimming in ponds, or hopping across a path or road on a rainy night. With their moist, permeable skin, frogs are never fully at home in a land environment with dry air and strong sunlight. They prefer humid conditions, and most are nocturnally active. Although the skin is a physical boundary, it is porous with respect to water. As a result, the water content of the frog's body can fluctuate strongly depending on outer conditions. A frog can lose over a third of its body mass through evaporation and still survive as long as it can replenish the lost fluid. Interestingly, frogs cannot drink through their mouths. Rather, they drink through their skin, especially their belly skin. A frog that is dehydrated can simply lie in a puddle and drink through its skin; or it can bury itself under leaves or in the soil and slowly draw moisture into itself. Desert frogs spend most of their lives up to a yard deep in self-dug burrows and slowly draw water out of the soil. Frogs can store large amounts of fluid in their bladders and distribute it as needed.

Frogs are dependent on warmth from their environment to maintain their body heat, so body temperature fluctuates with changes in ambient temperature. They are generally sluggish in cool weather, and some frogs can survive for a period of time in the frozen bottom of a pond. They become active in warmer weather, but you generally do not find amphibians basking in the sun as thick- and dry-skinned reptiles (think of lizards and snakes) do to warm up. They typically avoid the direct, evaporation-causing warmth of the sun.

So we see how the frog is very open to its environment. Through its skin it is giving up fluid to the air and drawing fluid in from the surroundings. Even though it has lungs, a frog still inhales around 40 percent of its oxygen and exhales more than two-thirds of its carbon dioxide through the skin. And the frog's body temperature oscillates with the warming and cooling of its environment. In these ways it lives in intimate connection, behaviorally and physiologically, with the changing conditions. Or we could also say the frog participates in these changing conditions and is part of them. There is no clear boundary indicating here the "frog" ends and there the "environment" begins. While we can say that the frog is a center of formative activity, this activity is wholly embedded within and dependent upon the larger fabric of interactions and substances that we call its environment. We can as little separate the frog from its environment as we can the center of a circle from its circumference.

As the name amphibian implies, frogs are beings between water and land. They are not wholly at home in water (as are fish) and are not fully at home on dry land (as are many reptiles). But they are not "homeless." They are at home in the in-between. They are aquatic for periods of time and when on land retain an affinity to moisture. In this sense they are "moist-earth" beings. This is even true of brightly colored tropical frogs that live high up in tree canopies—following a tendency of many tropical plants and animals to raise their "ground" into the crowns of trees. These frogs lay their eggs in little pools created in crevices or depressions of a tree or in rosettes of epiphytes, such as bromeliads, and stay mostly hidden from direct sunlight.

The frog's skin is moist and rich in glands. Some of the most potent animal poisons are produced in the skin of colorful tropical frogs. Poisons in reptiles or insects are usually created in glands within the organism. In frogs, the external organ of the skin maintains some characteristics of an internal organ—breathing, drinking, and secreting. From this perspective we can see how the so-called external environment of the frog in a sense belongs to or is part of the frog. This attunement is something you can sense almost viscerally in the early spring in the northeastern United States when the temperature rises and the first rains fall, and as part of this change the enchanting chorus of spring peepers and wood frogs fills the air.

Much of what I've discussed so far is true not only for frogs but for the other two groups of amphibians as well: salamanders and the little-known caecilians. What clearly sets frogs apart from these two groups is their form and the specific ways of behaving that are intimately connected with their unique body configuration. When a tadpole is reabsorbing its tail, the compact body takes shape as the head and body flatten and widen. The frog's short spine becomes quite rigid. The powerful rear legs grow longer than the body, as the drawing of a leaping frog vividly

FIGURE 8.2. A leaping frog about to land (*Rana esculenta*). (Adapted from Zisweiler 1976, p. 230)

illustrates (see Figure 8.2). This morphology and the saltational manner of movement is wholly different from that of its amphibian relatives.

Salamanders have a long body with relatively short legs. In some species the body elongates dramatically, while the legs become shorter and, in some cases, the rear legs do not develop at all. With a flexible spine and tail, they move with undulating, side-to-side movements of their body, which stays close to the ground. In fast movement, they hardly use their short limbs, whereas when moving slowly, they place their legs deliberately. The caecilians, which are tropical burrowing, wormlike amphibians, have no limbs and a very long body. In contrast to salamanders, they have no tails. With an extremely flexible spine, they push against the ground, and in alternation fold and extend their body to move along.

Amphibians form morphologically a spectrum with rich variations between the short-bodied, limb-dominated frogs at one pole and the long-bodied, limbless caecilians at the other (Figure 8.3). And while the dominant sense in frogs is sight, the caecilians are either nearly or fully blind.

The skeleton reveals in telling detail salient features of frogs. Frogs have the least vertebrae of any vertebrate, and the vertebral column (spine) is very short. Like all other amphibians, frogs have only one short neck vertebra, so that the head attaches almost without separation to the body. But the frog generally has only eight other vertebrae (some species have fewer) in its spine (including one sacral bone), while salamanders

FIGURE 8.3. Various amphibians; see text.

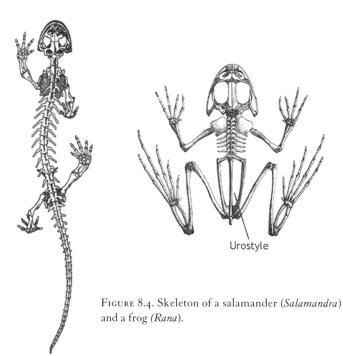

Urostyle

FIGURE 8.4. Skeleton of a salamander (*Salamandra*) and a frog *(Rana).*

generally have 15 to 20 (63 in the long-bodied siren). The skeleton of caecilians consists mostly of vertebrae—depending on the species between 95 and 285—and they have no tail.

Interestingly, while externally a frog has no tail, it does have one bone—the urostyle (or coccyx)—that corresponds to a tail in salamanders. This long bone develops out of three to four vertebrae that fuse together. It does not extend, however, beyond the pelvis; rather, it is drawn up into the pelvis and is a functional part of it (see Figure 8.4). Qualitatively this is a revealing characteristic. What would be part of the tail extending behind the body in salamanders is in the frog one long bone that is incorporated into the pelvis and helps to support and anchor the powerful rear legs. This detail expresses the overall contracted morphology of the frog's body that is correlated with the remarkable expansive development of the rear legs.

Now think of the way a frog moves. Sitting with its legs folded close to its body, the frog suddenly, like a spring, extends its legs, propelling it through the air. It cushions its landing with the forelegs and then the rear legs contract again at the sides of the body. Frog leaping is a radical kind of expansion and contraction that is morphologically mirrored in the compact body and the long, strong rear legs. Rapid, projectile-like movement also occurs in feeding when frogs use their "well-developed tongues [that] they are able to catapult from their mouths in order to pick up prey."[1] And when frogs croak, the body wall around the air-filled lungs contracts and forces air through the larynx, which suddenly relaxes and opens. The air streams over the vocal cords and into the mouth, filling the air sack, the skin of which vibrates. The surrounding environment fills with sound. The active animal expands out into the larger world. The chorus of many voices resounds in the spring landscape.

Do Frogs Come From Tadpoles?

Most of you have probably seen tadpoles in ponds and vernal pools.[2] With a thick squat body that abruptly tapers to a long finned tail, a tadpole can be distinguished from a fish. Yet, tadpoles are fishlike in many of their characteristics. They remain submerged in water and breathe through their skin and gills. They have no limbs and swim through water via undulating movements of a long boneless tail fin. Like fish, tadpoles have a lateral line organ, which runs along each side of the body and tail, through which they sense movements in water.

A tadpole typically grazes on algae that grow on plants, rocks, or at the surface of the water. Tadpoles have a "beak" and rows of denticles in their mouth that function like rasps to scrape off the algae. The denticles do not consist of bone and enamel but of keratin—a protein substance that, for instance, also makes up our fingernails and hair.

The tongueless tadpole sucks the algae into its throat, and the food enters the long intestine, where it is digested. There is no stomach. The intestine can be more than ten times longer than the tadpole itself and is its largest internal organ, making up over half of its body mass. Tightly coiled, the intestine takes up about half the space within the tadpole's ovoid-shaped body and is visible through the translucent belly skin.

How long a tadpole lives before it metamorphoses into a frog is dependent on the species and on outer conditions. A wood frog tadpole (*Rana sylvatica*), for example, usually metamorphoses into a froglet within two or three months after hatching in the northeastern United

Figure 8.5. Metamorphosis of the European common frog (*Rana temporaria*). Pictured to scale. (Photos by Tim Hunt, reprinted with permission; http://www.timhuntphotography.co.uk/)

Figure 8.6. Tadpoles (European common frog; *Rana temporaria*). (Photo by Friedrich Böhringer; Wikimedia Commons)

FIGURE 8.7. Northern leopard frog tadpole (*Rana pipiens*) viewed from below
(ventral; developmental stage 25 as given by Witschi). Note the coiled intestine
visible through the belly skin. (Adapted from Witschi 1956, p. 80)

States. The time is shorter when there are higher water temperatures
and ample food, and longer when there is colder water and little food.
Bullfrog tadpoles (*Rana catesbeina*) grow large—often around four
inches (10 cm) long—and, depending on circumstances, can sometimes
metamorphose into frogs in the fall (four to five months after hatching).
But, more typically, they live as tadpoles for two to three seasons before
metamorphosing.

If you would observe, side-by-side, a tadpole and an adult frog, you
would have no idea that the two animals have any connection with each
other. The fully aquatic, herbivorous tadpole bears no resemblance to
the tailless, four-legged, carnivorous, croaking and leaping frog. And
yet the two are inextricably connected; the one cannot exist without the
other.

The first external sign that a tadpole will not always remain a tadpole
appears in the gradual development of hind limbs. They originate as little
buds from the rear of the torso, grow into paddle-like structures, and then
elongate into muscular articulated limbs at the base of the still-existing tail.
While the hind limbs grow, the tadpole also grows, and the tail remains the
primary means of locomotion. The legs come fully into action only after
metamorphosis is completed.

While the hind limbs develop over many weeks or even months
(a gradual transitional phase is often called "prometamorphosis"),
the further transformation of tadpole into frog occurs within a short
period of time—often a week. Virtually nothing in the tadpole remains
untouched—organs and body parts are wholly broken down and

disappear, others are refashioned, and wholly new organs and body parts arise. While it is easy to say that everything changes, we gain a much richer sense of what such a transformation entails when we look at it in more detail.

In this case, it is not the devil that is in the details; it is the beauty and awe-inspiring transformative ability of life itself.[3]

Externally, the most marked transformation is the disappearance of the tadpole's tail and the concurrent rapid development of the forelimbs and growth of the hind limbs. The tail does not fall off. Rather, all its skin, muscle, cartilage, blood vessels, and nerves are internally broken down. The substances arising out of the self-digestion of the tail can be transformed and used to build up new body parts. Being tailless, a young frog is at first considerably smaller than the tadpole was. Depending on the species, it will remain small or will grow larger than the tadpole.

A tadpole breathes mainly by taking in oxygen through the thin and highly vascularized skin of its tail. The skin has been compared to fetal skin in mammals. The gills play a lesser role in respiration. Already prior to metamorphosis the tadpole begins to develop lungs and in some species—especially when the water is warm and stagnant—you can see tadpoles swimming to the surface to gulp air into their lungs.

During metamorphosis, while the tail is shrinking in size, the skin of the remaining tadpole thickens. It develops a wholly new pattern of pigmentation and a variety of secretory glands, some of which keep the skin moist once the frog leaves water to live on land. A frog's skin becomes less able to take in oxygen, and the lungs develop rapidly into the main organ of respiration while the gills are being broken down. Froglets begin floating near the surface with their nasal openings (nares) just above the water surface to take in air.

The circulatory system is intimately connected with respiration and experiences radical remodeling. All the vessels that serve the tail and gills are reabsorbed, and new vessels are formed connecting the lungs to the heart. The blood itself becomes thicker in consistency as more serum proteins are formed. Larval red blood cells—formed in the kidneys and liver—die off as the smaller and more numerous adult red blood cells are generated. In frogs most of the blood arises out of stem cells in the bone marrow. Different types of hemoglobin—which binds oxygen in red

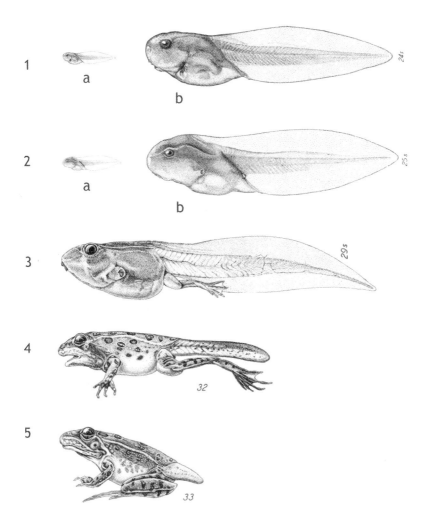

Figure 8.8. Tadpole metamorphosing into frog (Northern leopard frog; *Rana pipiens*). 1a, 2a, 3, 4 & 5 are drawn to scale, about 1.2 times natural size. 1b and 2b are enlargements of 1a and 2a respectively. Numbers next to drawings indicate the developmental stages as given by Witschi. (Adapted from Witschi 1956, p. 80–1)

blood cells—arise and they bind more oxygen than larval hemoglobins. While tadpoles graze mainly on plant matter, frogs feed on other animals, often insects. This transition means a massive remodeling of its feeding and digestive organs. The tadpole's beak is shed, its denticles are reabsorbed, and the mouth as a whole widens. A highly articulated jaw allows the mouth to open wide, and in many frogs true teeth form. In the mouth, secretory glands develop, as does a tongue, which is muscular and can be quite long. A frog often captures its prey by flipping out its tongue, to which the prey sticks; the frog then pulls it into its mouth, holds it momentarily with its teeth, and then swallows it whole.

FIGURE 8.9. Changes in the shape of the head during metamorphosis (*Bufo valliceps*; Gosner developmental stages 43, 44, & 45). Note the widening mouth. (Redrawn after McDiarmid and Altig 1999, p. 11)

Many herbivores have long intestines in which they digest their food, and this is the case in tadpoles, as mentioned above. During metamorphosis, three-quarters of the intestine degenerates, and the inner lining of the remaining intestine thickens; many folds arise in it, and a very large absorptive surface is created. As the intestine shrinks, a true stomach is formed, which secretes pepsin, an enzyme important for digesting proteins. While the rapid transformation of the digestive system occurs, the tadpole-becoming-frog hardly feeds.

If you picture a frog leaping to catch a mosquito flying by or a caterpillar inching its way up a stem and contrast this with the image of a tadpole scraping algae from a submerged stem, you have a sense of two very different ways of being and of relating to the surrounding world. This contrasting relation corresponds to a reorganization of the senses and sense organs during metamorphosis. The small sideways-directed eyes of the tadpole grow into large bulging eyes that let many frogs have a 360-degree field of vision and the ability to focus both eyes on one object. They gain the ability to move their eyes through the development of large external eye muscles. Eyelids allow frogs to open and close their eyes, which are kept moist by newly developed tear glands and ducts. Not only do the eyes grow, but their inner structure and physiology changes. For example, the spherical tadpole lens flattens, the double cornea fuses into a single cornea, and in the light-sensitive retina rhodopsin becomes the dominant pigment that reacts to light, as it is in most terrestrial vertebrates (and also marine fish).

Both male and female frogs can produce sounds and have a larynx with vocal chords that is not present in the silent tadpoles. Males are the dominant vocalizers in frogs—they are the ones we hear croaking loudly during the spring—and they have, in contrast to females, a vocal sac, an outpocketing of the floor of the mouth that fills with air and serves as a resonating body when the frog produces its sounds.

Anyone who hears a chorus of frogs can realize that frogs must have an acute sense of hearing. An eardrum develops in the frog that is flush with the outer skin and a middle ear that connects via a bony stirrup (stapes or columnella) with the inner ear, the only part of the ear that is developed in tadpoles. For a short time during this reconfiguration of the auditory organs the nascent frog is unable to hear sounds.

Manifold changes occur in other organs such as the brain, kidneys, liver, and pancreas. They are associated with the frog's different mode of perception, circulation, feeding, digestion, and movement. Therefore, as you can imagine by now, these organs also reconfigure both anatomically and physiologically.

The body of a tadpole is very flexible, and most of its skeleton consists of cartilage and not bone. As the frog develops, the bone formation increases. The limbs are fully developed after metamorphosis and the muscular hind limbs allow the frog at first to swim well with forceful rapid thrusts through the water and then to lead a leaping life on land.

Most frogs leave the aquatic environment and become land dwellers, although they thrive best in moist areas and often stay close to bodies of water. They return to water during the mating season. In the case of wood frogs (*Rana sylvatica*), for example, a female can lay over a thousand eggs, which are externally fertilized by sperm from the males. After fertilization the embryo begins to develop and forms into a tadpole that "hatches" out of its protective gelatinous ball and begins to live its tadpole life.

Thinking Development

In what follows I will be struggling with language. How can I adequately express what reveals itself during a study of amphibian metamorphosis? I don't want to fall back on standard phrasing that takes us away from the concrete richness and dynamism of what is showing itself. I want to stay close to the phenomena, not as a mere collection of facts, but as a transformative process. So bear with me. Try to catch my meaning. I'm trying to articulate something about development that usually gets overlooked.

In the process of metamorphosis a way of being we call "tadpole" disappears while a way of being we call "frog" emerges (and by frog here I mean the adult frog). No investigation of the tadpole alone could ever lead us to the knowledge that it will develop into a frog. The frog does not, in this sense, come from the tadpole. During metamorphosis an organic activity is at work that brings the tadpole to disappear while it brings the frog into appearance. We are witnessing a creative transformative activity as the frog becomes flesh—literally incarnates—during metamorphosis.

Development is not something automatic that just happens. What you find when you closely follow a developmental process is ongoing activity that cannot be accounted for by looking to what came before. You can't find the frog in the tadpole. This is self-evident as long as you attend to the actual process in its own terms.[4]

The ongoing creative transformation in development is beautifully described by T. H. Huxley, a colleague of Charles Darwin and one of the main early proponents of Darwin's theory of evolution:

> The student of Nature wonders the more and is astonished the less, the more conversant he becomes with her operations; but of all the perennial miracles she offers to his inspection, perhaps the most worthy of admiration is the development of a plant or of an animal from its embryo. Examine the recently laid egg of some common animal, such as a salamander or newt. It is a minute spheroid in which the best microscope will reveal nothing but a structureless sac, enclosing a glairy fluid, holding granules in suspension. But strange possibilities lie dormant in that semi-fluid globule. Let a moderate supply of warmth reach its watery cradle, and the plastic matter undergoes changes so rapid, yet so steady and purpose-like

in their succession, that one can only compare them to those oper-
ated by a skilled modeller upon a formless lump of clay. As with
an invisible trowel, the mass is divided and subdivided into smaller
and smaller portions, until it is reduced to an aggregation of gran-
ules not too large to build withal the finest fabrics of the nascent
organism. And, then, it is as if a delicate finger traced out the line
to be occupied by the spinal column, and moulded the contour of
the body; pinching up the head at one end, the tail at the other, and
fashioning flank and limb into due salamandrine proportions, in so
artistic a way, that, after watching the process hour by hour, one is
almost involuntarily possessed by the notion, that some more subtle
aid to vision than an achromatic, would show the hidden artist,
with his plan before him, striving with skilful manipulation to
perfect his work.[5]

It's interesting, and I believe significant, that Huxley is moved by the
phenomena themselves to reach for the metaphor of the hidden artist
sculpting the organism. Something creative—something I have referred
to as activity or agency—is molding the developmental process. But
it is not an artist creating something externally. It is the developing
organism as artist creating itself. This gives richer meaning to the term
"autopoiesis" (self-creation), which is often used to characterize the self-
organizing capacity of living beings.[6]

There is no need for dualism here. We do not need to think of some
being or life force that is somehow outside the process working in. We
just need to thoughtfully follow the process itself, and we see everywhere
organic life as "being-at-work"—Joe Sachs' felicitous translation of
Artistotle's term *energeia*.[7] The expression indicates that we don't have
two things: a being here and an activity there. The being is nothing other
than its working. It *is* only inasmuch as it is active.

But What About Genes and Hormones? Where Have all the Causes Gone?

I can imagine some readers are thinking, "That is all fine and good,
but there are physical causes such as genes and hormones that make
both tadpole and frog." We have been taught that science elaborates the
causes of things, and causes—so we imagine—always lie in the past.
Evidently, then, the cause of the development of the frog must lie in the

tadpole. Scientists start to investigate what substances—such as thyroid hormones—play a role in triggering the onset of metamorphosis and what genes are turned on and turned off while the tadpole is transforming.

There is always the assumption that some "thing" is the doer. The "thing" is primary and all activity is pictured to be the interacting of things (substances). In an abstract sense, the bare structure of DNA (the sequence nitrogenous bases) in a frog embryo, in a tadpole, and in an adult frog is, generally speaking, the same. If we begin by applying the widespread notion that genes consist of DNA, then if DNA stays the same, genes must stay the same. They are the stable and unchanging physical basis of the organism, while all other things may be different in the different life phases of the frog. If the genes are the "same" in embryo, tadpole, and adult frog, then can it be the genes that make these phases of life different from one another? This is worth pondering.

The conventional response would be: well, there are different genes that are acting at different times during development. So there's no problem; it's just that we don't know yet the total activation sequence of the ever-present DNA over time. But there is a problem, and it is hidden in the expression "genes acting." How do genes act? By being woven into the activity of the rest of the organism. A highly complex and variable series of interactions occurs when a gene "acts."[8] DNA is chemically modified (e.g., DNA methylation), brought into movement, repaired, and rearranged during the developmental process. To say that "DNA stays the same" is to say that certain structural features can be found to be stably produced and reproduced over time. That is basically the same as saying that over generations the wood frog stays a wood frog. When we say in biology something "stays the same," we actually mean it continually *becomes* the same out of activity; it is not an unchanging thing.

From a structural perspective, geneticists speak of about 25,000 protein-coding genes (specific sequences of DNA) in the human genome. But there are many more proteins than there are, in this static view, genes. About one million distinct proteins are estimated to be formed in the human body.[9] The synthesis of each of these proteins does require specific DNA sequences, but sequences are not simply lined up, waiting to be utilized. Their final specification occurs within the context of development and through the activity of the organism under changing inner and outer conditions. It has become clear, as several biologists stated in the article "How to Understand the Gene in the Twenty-First Century?" that genes need to be "conceived as emerging as processes at the level of the systems through which DNA sequences are interpreted, involving both the cellular and the supracellular environment. Thus, genes are not found in DNA itself, but built by the cell at a higher systemic level."[10]

At whatever level you consider—whether molecules (DNA, proteins, etc.), cells, tissues, or organs—you find interrelated activity. Surely, in the way science investigates, the doing will always be found in connection with "things," but the "things" don't explain the doings. DNA acts "because" proteins interact with them and act on them; proteins exist "because" DNA enables their synthesis. Every "actor" in the biological drama is also always an "acted upon." All the mind-boggling interactions molecular biologists discover *make sense* within the context of the healthy organism. They are part of the performance of the organism, to use neurologist and holistic scientist Kurt Goldstein's phrase.[11]

All the genes that "come into action" while the tail of a tadpole is being reabsorbed, or in the formation of the new type of hemoglobin in the nascent frog, are part of the unfolding story of the frog's coming into appearance. When scientists discover new molecular processes that in turn influence other processes, they are uncovering fascinating details about how the frog is coming into being. All this is interesting. The discoveries are a further elaboration of the development of the frog at a molecular level, but they are not "explaining" the frog.

It's an intriguing fact that at the molecular level processes and substances in very different kinds of organisms are remarkably similar. Many different organisms produce thyroid hormones, and, clearly, a boost in their production does not turn them into frogs! Researchers may say that thyroid hormone "controls" metamorphic changes in the tadpole-becoming-frog, evidently believing that discovering a substance that may normally be crucial in the realization of certain events is the same thing as understanding those events. You do not understand maturation of the skeletal system in human beings or the transformations in the tadpole-becoming-frog by studying only the associated substance-based conditions (hormones or genes). You have to study the human being or the frog.

In other words, the activity of hormones or genes can only be understood in the context of the given organism and the specific developmental process. This fact is vividly illustrated by development biologist, Scott Gilbert:

> The "meaning" of a particular molecule depends upon the cell receiving it. Consider, for instance, the wisdom of the frog: Don't regress your tail until you have already constructed your legs. How does the frog tadpole do it? One would think that the tadpole would use an early signaling molecule to cause the limb rudiments to proliferate and a later signal to tell the tail to regress. However, what happens is far more interesting—the same signal, the thyroid hormone T4 (thyroxin) is secreted by the thyroid gland. In some cells, the thyroxin signal is interpreted to say, "differentiate." In other cells, the thyroxin signal says, "die." And in others, the same molecule tells the cells, "proliferate." And in some tissues, the hormone doesn't seem to do a thing.[12]

The developing frog gives meaning to the substances it is employing. The significance of particular molecules and molecular events shows itself when we see them as an expression of the dynamic, orchestrating activity of the whole organism.[13]

Being Itself Differently

The tadpoles of a desert frog (*Spea multiplicatus,* called the desert or Mexican spadefoot toad) live in small ephemeral ponds in the southwestern United States and Mexico. The adult frogs live most of their lives underground in burrows and emerge only during the occasional rainstorms. When pools form, the females lay eggs; they are fertilized by the males' sperm, and tadpoles develop rapidly.

When they hatch, all tadpoles have the same basic morphology, but within a few days one can often find tadpoles with strikingly different morphology and behavior.[14] Many will be like a "typical" frog tadpole with small jaw muscles, smooth mouthparts, and long coiled intestines (Figure 8.10). They feed mostly on dead organic matter (detritus) and microorganisms, and they develop relatively slowly. But one can also find in the same pond larger tadpoles, which at first were thought to belong to a different species. They have large jaw muscles, notched and

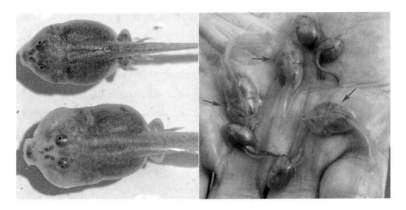

FIGURE 8.10. Two different morphs of tadpoles of the spadefoot toad (*Spea multiplicata*). Left panel: herbivorous morph (top) and carnivorous morph (bottom). Right panel: Both morphs on a hand; arrows point to the carnivorous morph. (Photos: David Pfenning; labs.bio.unc.edu/pfennig/LabSite/Photos.html)

serrated mouthparts, and a short, loosely coiled intestine. They develop this morph, as it is called, while feeding mostly on little fairy shrimp (and sometimes their fellow tadpoles). As carnivorous tadpoles, they develop a carnivorous body. Remarkably, they can transform back into the detritus-feeding morph if their food is altered. In some ponds one finds mainly the two different morphs, while in other ponds one finds mostly intermediate forms between the two extremes.

The desert ponds in which these tadpoles develop dry out rapidly— a long-duration pond exists for a month or so while a short-duration pond exists for only a couple of weeks. Interestingly, the fairy shrimp are most prevalent in short-duration ponds, and the carnivorous tadpoles are more likely to survive and metamorphose under these conditions, while in longer-duration ponds the omnivorous tadpoles are more likely to survive and metamorphose.[15]

Other environmental and maternal influences can affect the development of the carnivorous morph. For example, larger mothers tend to lay larger eggs and the tadpoles developing out of the fertilized eggs are more likely to develop the carnivorous morph. Or when tadpoles of a closely related species (*Spea bombifrons*) are also present in the pond, the latter tend to feed on shrimp and develop the carnivorous morph, while the Mexican spadefoot toad tends to feed on detritus.

So how a particular tadpole develops is intimately bound up with a variety of conditions that it meets in its environment and through its inheritance. The specific way these animals form and live depends largely on the active relation they establish with the environment, which in turn influences the formation and growth of their organs and body. Interestingly, you cannot tell the difference between frogs that developed out of the two different morphs of the Mexican spadefoot toad when they emerge from the pond.[16]

This is a vivid example of what biologists call phenotypic plasticity—the ability of an organism to develop, form, and behave in manifold and flexible ways in response to different and changing conditions in the environment. While spadefoot toads show remarkable plasticity as tadpoles, all organisms have phenotypic plasticity, some less, some more. Without plasticity organisms would be far too rigid and inflexible to navigate and thrive in the unpredictable real world.

Plasticity shows that every organism has the potency to be itself differently, to use philosopher Henri Bortoft's expression.[17] This potency is the capacity to become active that can manifest at any moment. When we hold this in mind, we realize that what I have called the active organism (being-at-work) is something much bigger than meets the eye. The "bigness" meant here is not something distinctly circumscribed that we can latch onto. As potency it is always more (and other) than all the particular realizations of activity that have previously manifested. It is not an extensive quality, but rather the intensive capacity of life itself.

In other words, the organism's potency is not some container full of specific capacities (a "tool-kit," as biologists like to put it) that it can put to work in response to new and changing circumstances. To imagine it in this way is to materialize, to make thing-like what is precisely not thing-like, since we are dealing with the actual active creating or bringing forth of new appearances. Thinking in terms of "genes" for plasticity would be misguided, which is not to say there are no substances or genetic relations that can support or constrain the realization of something new. We simply need to realize that even with the constraints and limitations that each organism has, it also has the ability to respond in an infinite variety of ways to new circumstances. The ways are not predefined. The organism's plasticity is its potency to come into action creatively. The potency is not a "what," a something we can point to, although this is unfortunately suggested by our use of a noun to suggest the reality.

From Space to Time, From Thing to Process

I want to counter the strong habit of thought that imagines the answer to developmental questions—the key that opens insights into the mystery of development—as lying in the past and in substances, in what one imagines to be physical causes. No matter at what level we consider an organism and its development, we are always dealing with organizing activity or agency that is specific and that provides the context for all part-processes, such as molecular events, that are discovered.

This is not to deny the contribution of the past to a developmental process. I just want to try to think of the relation in an adequate, close-to-the-phenomena way. The tadpole of a wood frog develops into a wood frog; the tadpole of a bullfrog develops into a bullfrog. That is the specificity that inheres in every aspect of a developmental process. In this sense what "is"—as coming from the past—makes possible and constrains what can become. But every "is"—when looking at the present moment—is also, in essence, activity.

I am encouraging a significant shift in attention. We habitually tend to consider an organism as "that which has become"—the organism as product that consists of a body, of heart and brain, of hormones and genes. We look at how these products are related and organized spatially and how they interact as products (this hormone affecting that organ). We conceive of everything as spatially bounded; we are tied to thing-ness in our minds, and the organism and its development appear in this light. Processes become the interactions of already existing substances, and development becomes the "chain reaction" sum of those interactions, the cascade of causal events. Time itself becomes "one thing after the other"—a sequence of events—and so also is atomized and made spatial.

Once we become aware of this grip of spatiality and thingness on our thinking, we can begin to loosen it when we attend to a developmental process. We attend to the process *as process*, follow closely the ongoing transformation, the coming-into-being and disappearing. We are no longer describing things, but flux and fluid movement. Huxley entered into this kind of attentiveness when observing the embryo, and he felt the need to characterize development as an artistic process. Development now shows itself as a true coming-into-being, the creative activity of life itself. Continuity lies no longer in the inertia of thingness but in the ongoing activity of life unfolding. This activity reveals itself as we move with the process in our thinking. As we observe, the continual flow in thinking is the means through which life-as-activity shows itself. In this mode of attentiveness, we no longer experience time as if from the outside, as a sequence of events, but rather as an ever-new-now, as ongoing creative activity.

This understanding of the organism and individual development (ontogeny) as creative activity opens up new territory and asks us to rethink all our notions based on a thing-centered, spatial way of viewing life processes. A biology of no-things—of activities—leads us into a science that takes seriously and strives to do justice to active, interpenetrating beings.

Creativity, Ancestors, and Origins: What Frog Evolution Can Teach Us

The perspectives on development and organismic life elaborated so far provide the basis for a fresh look at evolution. The history of life on earth shows, on the one hand, the stable existence of life forms over long periods of time and, on the other hand, the appearance of radically different life forms that never existed previously. The question is: How do we think about the relationship between what existed in the past and what arises as new organic forms during the course of evolution? Standard ways of stating the relation would be that all life ultimately evolved from bacteria-like organisms that lived billions of years ago, or, modern humans evolved from ape- or chimplike ancestors.

But what do these statements mean? The study of development in the present—as in the example of the metamorphosis of tadpole into frog—can sensitize us to the difficulties buried in such statements. Just as we can rightfully say that a frog cannot be derived from a tadpole, we can understand that humans cannot be derived from earlier primates, mammals and birds from reptiles, reptiles from amphibians, or amphibians from fish. In other words, inasmuch as a search for "ancestors" of a given group is looking to find more than a temporal antecedent, and is looking to find an explanatory "source" or "origin" of a group in the fossil record, it is misguided. It is a search full of expectations that cannot be fulfilled.

A central aim here is to unpack this perspective and to consider the insights and questions it leads to.

The Double Nature of Life

The study of development and organismic life in the present helps us to break through the habit of thought that only looks to the past to illuminate what develops and becomes active in any given present. The orientation to the past says, of course, that a frog comes from a tadpole. In so doing we are looking at the feature of continuity in space and time, of continuity in the sense world. We can always point to something that is present, formed, and alive "out of which" a next phase of life can develop. And this something is clearly important, because without it nothing could develop further, and also because there is specificity connected with it: Out of a salamander embryo a salamander develops, and out of a frog embryo a frog develops.

What exists in this way in the present is embedded in a long history. There have been thousands upon thousands of generations of wood frogs and bullfrogs. Over time, the life history of a given species is repeated—with variations of course—again and again. There is in this sense remarkable stability, with species staying more or less the same for generations.

This is all true. But it also is incomplete and one-sided. It does not encompass a central feature of development and organismic life that comes into view only when we look at the same phenomena differently. For this we need to shift away from focusing on how what already "is" provides the basis for what becomes. Our emphasis is no longer the causal approach that reigns today in the biological sciences and focuses on how the past determines the present. Of course, what is antecedent makes possible and also constrains future development. However, it does not provide insight into the special characteristics that arise during development and that make an adult frog so different from the tadpole.

The shift in perspective begins when we follow development and organismic life as process and transformation. When we stay in the flow of process itself, and notice the quality of changes that occur, something new shows itself. Instead of focusing on the past as determinant, we see ongoing creation.

It is in this sense that we can accurately say that the frog *does not* come from or develop out of the tadpole. You cannot study the tadpole alone and gain the knowledge that it will develop into a frog. In each generation "adult frog" comes into being through breaking down "tadpole." When developed, the adult frog actively maintains itself. From this perspective the organism shows itself as creative activity, agency, or being-at-work. The terms we use are not so important; what is important is that we perceive and become vividly aware of the creative, doing nature of the organism.

When we bring these two perspectives together, we see that life plays itself out in the polarity between creative activity and what has been created. (We could also say: between becoming and what has become, or between producing and what has been produced.) And, as in any true polarity, you don't find one pole without the other; they are not opposites that can exist separately. In a living organism, we observe what has been created (form, substances, structures) and we find creative agency. And just as all forms and structures emerge out of the creative activity of the organism, so is the creative agency always at work in what has already become. We are always dealing with formed life and the formative activity of life.

This double aspect of life is important. The formed life brings a kind of stability and constancy. We would not have anatomy, nor could we identify and classify groups if there were not some form of constancy and stability. At whatever level you consider the organism—its DNA, bones, or the countless generations of wood frogs that are always identifiable as wood frogs—there is stability.

FIGURE 8.11. Wood frog (*Rana sylvatica*): tadpoles (left) and adult (right).

 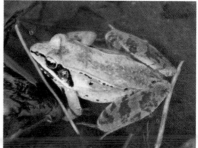

But it is essential to realize that such stability is not a static "is." Every form, structure, or substance is always being actively brought forth and actively maintained. This is being-at-work. All products have been produced; every creation has been created. The active, creative agency of life is always present, not as a thing but as the process of transformation—of coming into appearance and vanishing again.

Every enduring organic structure, every way of being, is a dynamic and persistent pattern in time and space—a pattern actively and creatively brought forth at every moment. This is the case whether we are speaking of the form of the eye in a single frog or the overall characteristics of a given species.

Fossils and Life Past: Portraying or Explaining

In fossils the earth preserves traces in the present of life past. For over 200 years geologists and paleontologists have discovered these fossil traces and striven to decipher and make sense of them. Inasmuch as we can read the fossil record, we gain an opening into the past. The fossil record presents us with a picture of a great diversity in forms of life that have inhabited the earth. It also points to great transformation. Only a few organisms alive today, such as horseshoe crabs, have been long present in the fossil record. Most existing species are much more recent appearances. Many groups, such as the dinosaurs, flourished for a span of time and then disappeared.

What the fossil record cannot show is a continuous process of transformation as we can observe it today in the development of individual organisms. We do not see a group of dinosaurs developing into birds in a continuous process, or a small fox-sized mammal morphing into today's horse. This is the limitation of a historical science. Fossils are tracks that can only point to a process of transformation. So, in principle, the fossil record leaves gaps; it reveals snapshots of life past. Moreover, most fossil remains are themselves incomplete and can only hint at the fuller reality of life past—usually only hard structures (teeth and bones in animals) and not soft vital organs are preserved, and of course all living behavior is absent. And the fossil record itself is also woefully incomplete. How few of the organisms that lived in the past actually left traces of their existence.

Figure 8.12. Fossil of a large ancient amphibian, *Sclerocephalus haeuseri*; from the Upper Carboniferous (Pennsylvanian) period in Germany. Body length: about 5–6 feet (1.5–-1.8 meters).

Fossils of frogs appear much later. (State Museum of Natural History, Stuttgart, Germany; Dr. Günter Bechly. https://commons.wikimedia .org/wiki/ File:Sclerocephalus_haeuseri,_original_fossil.jpg)

All this is good reason to tread gently when we interpret the fossil record and to remain cognizant of all we do not know.

One focus of evolutionary research is to collect, describe, and order the fossils found. This is in itself a huge undertaking. New fossils are continually discovered, which both enriches and makes more complex the emerging picture. Such descriptive and comparative paleontological work remains close to the phenomena themselves. There can be considerable disagreement in details about the significance of a given fossil and its taxonomic relations, or about the purported function of particular structures, but the fossil record clearly suggests that organismic life in all its diversity evolved over time as part of the evolution of the whole earth. And when one brings the fossil record into relation to all the knowledge one can gain by studying organisms, ecology, and geology in the present day, a rich picture emerges.

It is one thing to paint a picture of the phenomena that suggests evolution. It is another matter to set out to explain how evolution happened. This is a second focus of evolutionary research. Here scientists fill the

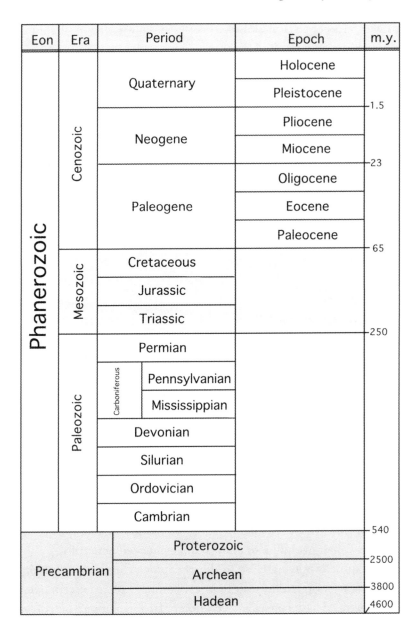

Eon	Era	Period		Epoch	m.y.
Phanerozoic	Cenozoic	Quaternary		Holocene	
				Pleistocene	1.5
		Neogene		Pliocene	
				Miocene	23
		Paleogene		Oligocene	
				Eocene	
				Paleocene	65
	Mesozoic	Cretaceous			
		Jurassic			
		Triassic			250
	Paleozoic	Permian			
		Carboniferous	Pennsylvanian		
			Mississippian		
		Devonian			
		Silurian			
		Ordovician			
		Cambrian			540
Precambrian		Proterozoic			2500
		Archean			3800
		Hadean			4600

TABLE 8.1. Geologic time periods—from most recent (top) to the oldest layers of rock (bottom); m.y. = millions of years (as estimated by measuring radioactive decay in the respective rock layers).

gaps with ideas of a special kind—theories and purported mechanisms. It is a highly speculative undertaking.

Thinking about the history of life in terms of evolution began in the late eighteenth and early nineteenth centuries. Since then, there have been numerous theories of evolution.[18] Charles Darwin's 1859 book *Origin of Species* provided a cogent theory that has dominated evolutionary speculation ever since—to the degree that most people today, when they hear "evolution," have in mind Darwin's theory of evolution, as though the two were synonymous.

In recent years, many tenets of Neo-Darwinism (the modern form of Darwinism) are being called into question by scientists and philosophers.[19] This indicates that there are clearly different ways of explaining evolution and that it is possible, at least to a degree, to separate the general theoretical framework from the phenomena scientists are attempting to explain.

I have always been interested in evolutionary phenomena and new findings, and at the same time have remained skeptical about speculative attempts to explain evolutionary processes in the past. Yet I am also not free of preconceptions and have had to work to free myself over time from ingrained habits of thought and assumptions that I didn't even know I had.

When I try to hold back the activity of "theorizing about" and "explaining," and instead apply the potency of thought to perceive what can reveal itself between the phenomena, sometimes I see that these phenomena can be understood in a very different way. Through this work my mind has opened to new possibilities of understanding, and I have been able to move beyond what I thought I knew. This is a process of ongoing inner evolution to understand evolution more fully.

In the spirit of staying close to the phenomena and using thinking as a means to see relations instead of explaining, I offer here an open-ended portrayal based on a consideration of frog evolution as it reveals itself through the fossil record. Given the caveats about the fossil record mentioned above, we have every reason to hold conclusions lightly and keep our picture of evolutionary processes open, mobile, and ready to evolve. Clearly, my aim is not to come up with any explanation or theory of evolution in the conventional sense. And I do not claim to be offering an exhaustive treatment of the topic. But I do want to shed some light on evolution that has been blocked by dominant habits of thought.

Frog Fossil History

Today there are about 4,800 known frog species. Each has its particular characteristics, and it is even possible to identify a species on the basis of a few bones. But all frogs have very similar skeletons, and since frog skeletal structure is unique among all four-legged vertebrates (tetrapods), a specialist examining fossil bones or imprints can identify whether they belonged to a frog or not.

For the most recent periods of the earth's history one finds fossils that can easily be identified as frogs. For example, wood frog fossils have been found in layers of rock in Nebraska (Pliocene epoch of the Cenozoic era) in which fossils of now extinct animals such as sabertooth cats and stegomastodons (relatives of elephants) are also found. Frog fossils can be found back into the Mesozoic era (colloquially known as the age of dinosaurs), some very well preserved (see Figure 8.13). In some cases one even finds fossils of tadpoles and partially metamorphosed frogs in one layer.[20]

Figure 8.14 shows the skeleton of a modern frog (a) and a reconstruction of the skeleton of one of the earliest frog fossils found until now (b). This fairly complete fossil, given the name

FIGURE 8.13. Well-preserved fossil frog skeleton (*Liaobatrachus*) found in the early Cretaceous period in China. (From Roček et al. 2012)

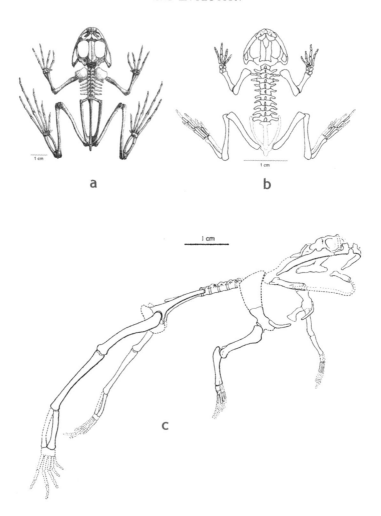

FIGURE 8.14. a: Modern frog skeleton (common European water frog, *Pelophylax esculentus*); b: reconstruction of a fossil frog (*Viaraella herbsti*) from Argentina, middle Jurassic period; c: reconstruction, shown as if jumping, of the currently earliest known fossil frog, *Prosalirus bitis*, from Arizona, early Jurassic period. (From, a: https://commons.wikimedia.org/wiki/File:Rana_skeleton.png; b: Roček 2000, p. 1301; c: Shubin and Jenkins, 1995)

Viaraella herbsti, was found in Argentina, and all the bones resemble those of modern frogs. The earliest species found up until now that is considered a frog, *Prosalirus bitis*, was discovered in Arizona (three specimens from early Jurassic period of the Mesozoic era; Figure 8.14c).[21] With its long hind limbs, lengthened pelvis, the presence of the urostyle (which is unique to frogs), and short body, it is clearly a frog. It does have some characteristics in the skull and other parts of the skeleton that distinguish it from modern frogs, but not to the degree that one would think it to be a different kind of animal.

So it seems that since the early Jurassic period of the Mesozoic era, the basic frog way-of-being, at least as it is manifest in the skeleton, has hardly changed. A paleontologist who specializes in amphibian evolution writes that, "the basic structural scheme of frogs has been maintained without any significant change, which suggests that an equilibrium between function and structure and the mode of life was maintained."[22]

The earliest known salamander and caecilian fossils—the other two groups of living amphibians—are also found in Jurassic layers.

No frog fossils have been found that date to before the early Jurassic period. In older rocks (early Mesozoic and Paleozoic eras back to the Devonian period) one finds many fossils of amphibians, all of which have long been extinct. Both larval and adult fossils have been found, mostly in rock formations that geologists believe formed out of sediments at the bottom of ponds and lakes. While in many cases the fossils superficially resemble today's salamanders, they also have their own array of characteristics that set them apart from the living groups of amphibians (frogs, salamanders, and caecilians).

Most of these early amphibian fossils were much larger than today's amphibians, ranging from 3 to 18 feet (1 to 6 m) in length. There is an astounding diversity of forms, as if nature were experimenting with manifold ways to be an amphibian (see Figure 8.15). But it is evident than none resemble a frog. As the eminent paleontologist Robert Carroll writes, "Despite the great diversity of Paleozoic and early Mesozoic tetrapods that had an amphibious life history, none shows obvious affinities with the three living amphibian orders. This constitutes one of the largest morphological and phylogenetic gaps in the history of terrestrial vertebrates."[23]

Such "gaps" are popularly called missing links, and paleontologists have been motivated to continue to search for fossils that would fill the gaps. There is an expectation of some form of continuum—fossils that

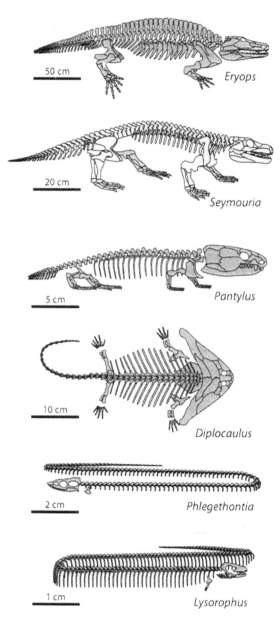

FIGURE 8.15. Some examples of the diverse types of amphibian fossils that have been found in the early Mesozoic and Paleozoic eras, before any fossils of the living groups of amphibians (frogs, salamanders, and caecilians) are found. Not ordered temporally. (Adapted from Schoch 2009)

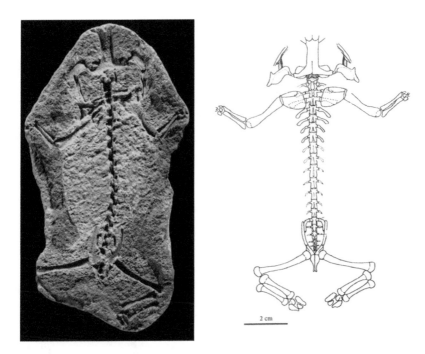

FIGURE 8.16. *Triadobatrachus massinoti*; amphibian fossil from the early Triassic period. Left: fossil imprint; right: reconstruction. (Imprint from Musee d'Histoire Naturelle, Paris; reconstruction from Roček and Rage 2000)

reveal a transition from older amphibians to frogs (and also to salamanders and caecilians). This expectation is based on the view—which Darwin presented in 1859 so forcefully and cogently in *Origin of Species*— that species evolve gradually out of one another. So paleontologists hope to find fossils that display at least some frog characteristics to bridge the gap between full-fledged frogs and early amphibians. While this hope motivates the search for fossils, it does not generate them, so paleontologists have to live with what they find.

In the case of frogs, a very few connecting fossils have been found. One is *Triadobatrachus massinoti* (from the early Triassic period of the Mesozoic era; see Figure 8.16). Only one specimen has been discovered so far, and that was in Madagascar. The skull is quite froglike in overall shape and in the configuration of the individual skull bones. In contrast, the body is not froglike: It has a relatively long vertebral column, ribs, a tail, and short hind limbs. In other words, the specializations connected with the present-day frog's leaping mode of locomotion were not present. There is some lengthening of the pelvis and the vertebral column is shorter than in many other amphibians, so the fossil is morphologically "an intermediate between primitive amphibians and anurans [frogs]."[24]

Pelvic lengthening and characteristics of the sacrum in another fossil, *Czatkobatrachus polonicus*, show greater resemblance to frogs. But no skull bones have been found, and otherwise it shows little resemblance to frogs. *Czatkobatrachus* was found in Poland and in somewhat younger layers than *Triadobatrachus*.[25]

Some substantially older fossils from the lower Permian period of the Paleozoic era do not resemble frogs, but they do have a few froglike characteristics that only a paleontologist with highly specialized knowledge would recognize. One example is in *Doleserpeton annectens* (from Oklahoma, Figure 8.17). Its most froglike characteristics, as in *Triadobatrachus*, are in the skull. As is often the case, the teeth—the hardest parts of an animal—are well preserved and have a definite morphology. *Doleserpeton* has so-called pedicellate teeth, in which both the crown and base of each tooth are made of dentine and the two parts are separated by a narrow layer of uncalcified dentine. This morphology is characteristic of frogs and other modern amphibians, but not of most extinct groups. Also froglike are the structure of the palate, the stirrup (stapes) in the middle ear, the inner ear, and features of the braincase. Otherwise it resembles other extinct four-legged creeping amphibians and is more like a salamander in overall form than a frog.

Gerobatrachus hottoni (from Texas, Figure 8.17) has, in contrast, a more compact form. It has a broader, more froglike skull shape and also a shorter spine.[26] It too has pedicellate teeth. Both of these species, unlike the many larger ancient amphibians, are in the size range of modern amphibians.

So in these older layers there are traces of "frogness." But they are present in different species, each of which is uniquely configured. Frogness with all its features does not appear all at once, nor does it appear in only one lineage in the fossil record.

Going back still further, paleontologists continue to find amphibian-like animals. They represent the first four-legged (tetrapod) vertebrates in the fossil record. Some of the earliest tetrapod vertebrates, such as

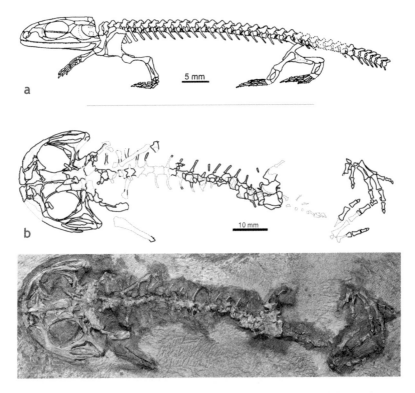

FIGURE 8.17. a: *Doleserpeton annectens*; amphibian fossil found in Oklahoma, lower Permian period. Body length: approx. 5.5 cm (2.17 inches); b: Partial reconstruction and photo of fossil of *Gerobatrachus hottoni,* amphibian fossil from the lower Permian period, found in Texas. Body length: approx. 11 cm (4.3 inches). (From, a: Sigurdsen and Bolt 2010; b: Anderson et al. 2008)

Acanthostega (see Figure 8.18), also have fishlike characteristics. These include bony fin rays in the tail, evidence of a lateral line organ, and evidence of internal gills. As paleontologist Jennifer Clack writes, "If one were to imagine a transitional form between a 'fish' and a 'tetrapod,' *Acanthostega* would match almost exactly those expectations."[27] The detailed anatomy of the four limbs does not suggest that those limbs could support walking on land. More likely, they were used as swimming paddles, which indicates that the seemingly logical and often-presented notion that four-leggedness (tetrapod limbs) in vertebrates developed as an adaptation to living on land isn't valid. Tetrapod limbs developed first in water and were used for moving on land only in later forms.

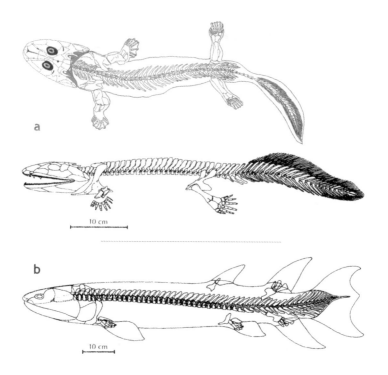

FIGURE 8.18. a: *Acanthostega*, a fossil from the late Devonian period that exhibits both amphibian and fish characteristics; see text. b: *Eusthenopteron*, a lobed-fin fish fossil from the Devonian period; see text. (From: a, top: Clack 2012, p. 165; a, bottom and b: Carroll 1997, p. 300)

The lobed-fin fish *Eusthenopteron* is a vertebrate fossil from the Devonian period predating any tetrapods (see Figure 8.18). Although clearly an aquatic-dwelling fish, it has certain structures in common with the tetrapods that arise later. Most striking is the internal structure of the fins. The body-near parts of both the pectoral and pelvic fins consist of three bones each. They correspond, in the pectoral fin, to the humerus, ulna, and radius in the forelimb of a tetrapod and, in the pelvic fin, to the femur, tibia, and fibula of the hind limb of a tetrapod. The arrangement of some of the skull bones as well as the presence of internal nostrils are also similar to subsequent tetrapods.

Fishes represent the first vertebrate animals to appear in the fossil record. This takes us far enough back for our purposes here—back to the "age of fishes," when manifold types of fishes populated the waters of the earth and no fossils of amphibians, reptiles, birds, or mammals are found.

Forming a Picture of Evolutionary Transformation

From the fossil record we can know that frogs have been a creative presence on earth for a long time. We find frog fossils back to the early Jurassic period of the Mesozoic era. Many species have arisen and passed away since then, but frogness (the order Anura in scientific terms) has remained present. The diversity of frogs increased over time, and today's variety, as expressed in 4,800 species, shows the many wondrous ways of being a frog that have evolved.

In earlier layers there are no frog fossils. The first frog fossils have virtually the same proportions and the same skeletal morphology as today's frogs. So since the early Jurassic, the highly specialized and unique morphology of frogs has remained remarkably constant.

But there are also rare and interesting transitional forms. Before there are full-fledged frog fossils, *Triadobatrachus* and other fossils exhibit some few features—mostly ones in the head—that later all frogs possess. These animals were a far cry from frogs, but if you know frog morphology well, you can see hints of what is to come. Of course, you could never predict by knowing *Triadobatrachus* that frogs would appear later.

What is also typical in the fossil record is that the hints or foreshadowing of what will come later are not manifest in only one type of fossil, but in several. Various elements of what appears later in the new group are manifest in earlier periods, but in different lineages. Evolutionary scientists often speak in this connection of "mosaic" evolution, since various characteristics appear in different arrangements in different organisms.

In some groups of plants and animals paleontologists find more transitional forms than in others. But even when a trove of fossils is available, such as in the horse family (Equidae), it is not the case that they line up in a neat series. Rather, there is surprising diversity in the forms that predate modern horses.[28] Evolving features appear in different lineages. This is also vividly visible in human (hominid) fossil history—the more fossils that are found, the less straightforward the emerging picture of the evolving human form becomes.[29]

If we consider this feature of the fossil record from a bird's-eye perspective, it is as if we are seeing hints of what is to come spread out in various earlier forms, which then become extinct. Eventually new forms appear, sharing characteristics with various earlier forms but in a new configuration that could never have been predicted based on what came before.

In the Mesozoic and late Paleozoic eras there was a great diversity of amphibians, but only the relatively late appearing frogs, salamanders, and caecilians survived to the present. Among the extinct amphibians, paleontologists find the first four-legged vertebrates (tetrapods) in the fossil record. Just as amphibians today are beings that thrive at the interface of water and land, so the earliest amphibians were four-legged but lived mostly in water. These early tetrapods are preceded in the fossil record of vertebrates, as noted above, by a plethora of ancient fish. Hints of the tetrapod future can (in hindsight) be found in the group of lobed-fin fish. They possessed, as we have seen, a bone structure in the fins that can be viewed as the first beginnings of four-leggedness.

Such phenomena in the fossil record present a picture of major transformation. Manifold new ways of being in the world come to manifestation. This is the narrative of evolution. Today we find radical organic transformation in the individual development of organisms—in embryonic development or in metamorphosis like that from tadpole to frog. Today frogness "becomes flesh" in each generation of frogs. During the earth's history there was a span of time in which frogness became flesh (and sometimes fossilized in rock!) in the stream of then-existing vertebrate life. As individual development is an act of creation, so is evolutionary history a record of creativity writ large.

Over the long ages we have been discussing, evolution was occurring with myriad types of organisms. Different ways of being came into appearance—all the kinds of microorganisms, plants, and animals. Since every organism is connected with others in the realization of its life, we need to think of organisms as interpenetrating fields or centers of activity that are in turn influencing and being influenced by the whole ecology of the earth. The fossil record reveals that the earth is evolving as a whole, and it presents to us traces of a global process of creative transformation.

The Problem of Ancestors

In my portrayal of frog and amphibian evolution I have tried to present a picture of evolution as a creative process. In so doing I have consciously avoided a trend of thought and research that is often dominant in the study of the fossil record and evolution. That is the search for ancestors and origins in the fossil record.

A hope or expectation is engendered that if we just find the right fossils we will have the answer to questions like "Where do frogs come from?" or "What is the origin of human beings?" This is a form of reification, of mistakenly picturing entities as the full reality of much larger ideas and relations. The philosopher Alfred North Whitehead spoke of the "fallacy of misplaced concreteness."[30] In the case of origins, we are dealing with a riddle much broader and deeper than what particular fossils can tell us. Origins are not located "in" a tangible something. We only need to ponder the fact that knowing my family genealogy may tell me interesting and important things, but it definitely does not satisfy my quest to really know who I am and where I come from in a less tangible (but not, therefore, less real) way. If we narrow the search for origins to the fossil record, then we will either end up dissatisfied with what we find or we will live in the illusion that we have understood more than we have.

One way I have avoided this narrowing of perspective is to purposefully refrain from using certain forms of expression. I did not say: "Frogs evolved out of primitive amphibians." Or: "The first amphibians evolved out of lobed-fin fish." The phrase "out of" is not as innocuous as it first may seem. This becomes apparent when you see how "out of" is conflated with the notions of "ancestors" and "origins." We need to make distinctions here.

Let's start with "out of." I can observe today how a frog develops "out of" a tadpole. A frog does come from a tadpole inasmuch as we are looking at the continuity within development. There is no break. This fact finds expression in the more general phrase "life comes from life." Likewise, "out of" in connection with evolution points to an unbroken stream of life in the course of evolution. There has been physical continuity in the evolution of life, just as there is physical continuity from the tadpole to the frog. But the unbroken flow of life is not an entity; it is not like a fossil. It is what we conceive of as the potency of evolving life that continues through biological reproduction from generation to generation. Without conceiving of such potency and its continuity, we cannot even conceive of evolution. So when we say "out of" in relation to evolution, we need to restrict its meaning to indicate simply the physical continuity of life as a prerequisite for evolution.

"Out of," in this restricted sense, does not suggest that we can account for what appears as new in a developmental or evolutionary stream by investigating a physical source. This is what we saw in the metamorphosis of the tadpole into the frog. The frog is something new in relation to the tadpole; we can't understand the adult by examining the tadpole. To begin to understand this metamorphosis we need to see it as a truly transformational process. The creative, originative developmental power appears in the whole developmental stream, not in one particular moment, and is not accounted for by a physiological or genetic "trigger."

Similarly, antecedent amphibian fossils do not reveal a point of origin. They indicate the pathways through which frogness—what is new—comes to appearance in the fossil record. *Triadobatrachus* (or any other fossil species that may be found in the future) does not account for the later appearing frog; it shows us part of the story of an evolutionary process. In this sense, every new fossil illuminates—and often makes more complex—the emerging picture of the evolution of any given group of organisms.

In evolutionary paleontology and in paleoanthropology (dealing with fossils related to human evolution), there has been and continues to be a strong drive to find and identify a particular fossil species as the ancestor of a subsequent group. One hopes to find, say, the last common ancestor of humans and apes, or of frogs and salamanders. This ancestor would be the species "out of" which the later forms arise and there is the expectation that such fossils would answer questions about the origins of a particular group of organisms. We are dealing here with the conflation of "out of," "ancestors," and "origins."

Just from studying the research, you can discover that something is awry in this undertaking. A first indicator is that nowhere in the research reported in the primary literature do you find the explicit claim: we have found fossil species x that is the ancestor of fossil species y and z. One looks for ancestors and discovers often a plethora of new fossils that in some cases have characteristics no one expected. But no one ever finds a species that is *the* ancestor of more recent ones. This is an interesting co-nundrum and reveals a deeper problem in the search for ancestors.

Consider a couple of statements about fossil ancestors that lead us further:

> Most paleontologists look for ancestors—an ancestor-descendant sequence in which ancestors are assumed to be generalized in a particular character, and the descendants more specialized.[31]

> An "ancestor is, by definition, plesiomorphic (primitive) in every way relative to its descendants."[32]

The ancestor is understood to be more "generalized," "primitive," or in technical language, "plesiomorphic" because, logically, it cannot already have the characteristics of later forms. The ancestor cannot already have the specialized characteristics that define its different descendants.

But there is a problem here, which is vividly alluded to by Alfred Romer, a great twentieth-century paleontologist: "After all an animal cannot spend its time being a generalized ancestor; it must be fit for the environment in which it lives, and be constantly and variably adapted to it."[33] All fossils reveal variously specialized organisms.

If a fossil has characteristics of frogs already, then it is not the ancestor of both frogs and salamanders, since it is already showing frogness. But if a fossil shows no frogness, then how should we determine whether it is a frog ancestor or the ancestor of some other creature? It can't be done. As vertebrate paleontologist Robert Carroll writes, "If all relationships are established by the recognition of shared derived [i.e. specialized] features, ancestors cannot be recognized as such because they lack derived traits that are otherwise thought to characterize the group in question."[34]

This glaring problem has not stopped the search for ancestors. What happens in practice, however, is that scientists usually end up recognizing that purported missing links or ancestors are too specialized in one way or another to fit the vaguely held notion or expectation of an ancestor. For example, Romer designates an apparent prime candidate for a tetrapod ancestor as being "a bit off the main line" due to its unique specializations.[35]

In 2009, a new hominid fossil species—*Ardipithecus ramidus*—was described in many articles in *Science*.[36] The fossil fragments, which evidently stem from many different individuals, reveal a species "so rife with anatomical surprises that no one could have imagined it without direct fossil evidence."[37] It is a fascinating discovery because it differs starkly from prevailing suppositions about how a common ancestor of chimps and humans should look. Most researchers have pictured this ancestor as essentially a chimpanzee-like animal (although there have long been dissenters from this view[38]). This is the basis of the talk about "humans evolving from apes." Then *Ardipithecus* is found, and it turns out to be wholly its own type of creature that no one would have predicted. To mention just one puzzle: its morphology indicates that it was not a knuckle-walker, like living great apes, and that it could also walk upright; yet its brain is the size of a chimpanzee's. A great and fascinating surprise—one more riddle in the diversity of life forms. This is the kind of discovery that happens in all areas of fossil research.

Because it is an early hominid fossil (predating the famous Lucy [*Australopithecus afarensis*]), the authors maintain that *Ardipithecus* "provides the first substantial body of fossil evidence that temporally and anatomically extends our knowledge of what the last common ancestor we shared with chimpanzees was like."[39] So now *Ardipithecus* is being viewed by many as a new kind of model or framework to better picture this purported ancestor. Don't they expect that they may well find more "strange" fossils that don't fit their mental models? Don't they see that new fossils continually widen the picture rather than pin it down?

Due to the age and morphological characteristics of *Ardipithecus*, its discoverers write in a more recent (2015) article, that the species is "obviously not the common ancestor of humans and chimpanzees."[40] But this does not keep them from claiming that it "nonetheless provides strong evidence with which to infer that the [common ancestor] was a generalized African ape." But isn't that just one more vague mental construct? Doesn't *Ardipithecus* reveal its own suite of specializations? And, as we heard above, if, by some wonder, something "generalized" were found, no one would be able to determine whose ancestor it might be, since it would be lacking in characteristics of the later forms.

So, it appears that the "last common ancestor," which is supposed to be a fossil species, is actually a speculative mental construct that is a placeholder for the particular species or group from which later forms would be assumed to have evolved. When such a speculative construct is used as a model to evaluate the status of actual fossils, it clearly biases that evaluation. The chimp model of human ancestry has biased research for well over a hundred years. Wouldn't it be much more scientific to just renounce the drive for reification and for using models based on extant or fossil forms, and wait and see what one finds?[41]

When our focus is finding what we believe to be origins in the past, then the danger arises that we view what is new as essentially a reshuffling or a reordering of what already existed. Birds become "glorified dinosaurs" and humans the "naked ape" or "third ape." We denigrate the new that arises in evolution. We miss what is creative in evolution.

Of Origins and Creative Evolution

When we consider present-day or evolutionary development, we can use the expression "out of" to point to the unbroken continuity of life: all life forms "come from" previous life forms. We can say humans "evolved out of" the evolutional lineage that also brought forth monkeys and apes. But we could also say the "humanness manifested into" this lineage, a way of expressing that may sound a bit strange, but points to the fact that new characteristics cannot be derived from past forms. New characteristics that emerge in the stream of evolution cannot be explained by the past.

The necessary counterbalance to the focus on the past as the precondition for what arises later is the focus on the new in its own right. This entails recognition of the special characteristics, in their integrated wholeness, of new forms that never before existed. Then we begin to get a glimpse of the truly creative or originative, both in developmental processes today and in evolution.

At any stage of life an organism is both past (what has become) and activity that brings forth something new. The organism as activity or agency is not some entity in space, some trait or characteristic that you could place next to its skin or stomach. It is the being-at-work in all the features of the organism. It is not something we can directly perceive as an entity. What is creative in development shows itself to the mind's eye as we follow a developmental process from embryo to adult. It shows itself when we study the way an organism manifests plasticity by "being itself differently" as conditions vary. It shows itself when we compare one type of organism with others and begin to see its special way of being reflected in all its features. (Remember the comparison of frogs with salamanders and caecilians).

Similarly, the creative and originative in evolution begins to show itself in the fossil record when we build up a picture of the temporal, morphological, geographical, and geological relations. We can see, not the steps as in a neat progression, but the wondrous variety of forms that let us intimate, say, the way frogness or humanness has come to expression in the world.

When we bear in mind the vast diversity of life forms on earth in the past and present, this is a daunting task. But it is also a stimulating open-ended exploration that becomes enriched by every new fossil and by every new insight into connections.[42]

So when we think of discovering origins, we actually need to look to the whole of creation, not to assumed fossil ancestors.

*   *   *

With the evolutionary appearance of humans on earth and subsequent historical and cultural evolution, beings have arisen who are in a position to consider their own evolution and the evolution of the whole planet. This is a new thing under the sun. We can study and ponder the evolution of all our fellow creatures on earth, and of the earth itself. This is a characteristic that we do not find in the rest of organic life. It is a unique quality of human evolution that we arise as beings who can study and begin to understand, through thoughtful observation and contemplation, the evolving world of which we are a part and in which we participate.

When we study evolution we are consciously connecting with the whole of life—the life with which we also are connected through evolution. In this sense, evolution is reflecting back on itself in the minds of human beings. But this reflection is itself a creative activity; it is not a given. The more I study evolution, the more I see the boundaries we put in the way of an expansive and deeper understanding. But I also see that we can move beyond those boundaries.

It becomes ever clearer that our understanding of evolution will evolve to the degree that we evolve in our capacity to see evolution as a creative activity.

PART III

Taking Responsibility

CHAPTER 9

The Dairy Cow
and Our Responsibility
to Domesticated Animals

IN CONTRAST TO THE OTHER ANIMALS you have learned about in this book, the dairy cow is a domesticated animal, one whose past, present, and future—down into the core of its biological makeup—are directly and inextricably connected with human activity. Through thousands of years of interchange we have become part of a cow's being, and she part of ours in a way that goes beyond the connection we have with wild animals. Cows are deeply dependent on us and we on them. This bond makes the question of what responsibility we have to cows (and to all domesticated

FIGURE 9.1. A grazing dairy cow from Hawthorne Valley Farm, Ghent, NY.

animals) loom large. How do we view this relationship, and how does that view guide our intentions in the way we breed and treat these animals? Do we see cows as beings who serve us and for whom we respectfully care? Do we see them as units of production whose efficiency we need to maximize? Do we manipulate them as bioreactors to produce substances we desire? You can find all of these perspectives expressed today, and they all have consequences.

Out of the Life of the Dairy Cow

Cows are grazers, like their wild cousins the bison. If they are allowed to live a life that corresponds to their nature, they live on pastures—in the midst of the food they eat—grazing on grasses and wildflowers. The cow lowers her head to the ground and touches the plants with the front end of her soft, moist snout. She does not chomp off the plants with her teeth. In fact, the cow (like the bison, giraffe, and other ruminants) has no top incisors or canines. She has, instead, a tough fibrous dental pad at the front of the hard palate. When feeding, the cow reaches out with her rough, muscular tongue, enwraps the plants, and tears them off while slightly throwing her head upward and to the side. She clearly needs to use her tongue for feeding—cattle that receive soft, fiber-poor feed begin to lick their fellow cows much more than usual to compensate for the lack of interaction with the tough, fibrous grasses and forbs. The cow needs this interaction to remain healthy.

Taking about one bite per second, the cow moves slowly through the pasture. Large glands secrete saliva while she grazes, and after taking many bites she swallows the now moistened mass of food. She can continue grazing in a kind of flowing rhythmic persistence for a couple of hours at a time. Cows on the pasture usually have a number of such feeding periods during the 24-hour-day, spending about one-third of the total day grazing. When swallowed, the food reaches the rumen, the huge first chamber of the four-chambered stomach. Occupying the entire left side of the abdominal cavity, the rumen can hold up to 45 gallons of fluid and feed. The muscular rumen massages the food in regular contractions—about one to two per minute is a sign of a healthy cow. It is only when a calf begins to feed on grass that the rumen completes its development and becomes fully functional. You could say that grass is the environmental

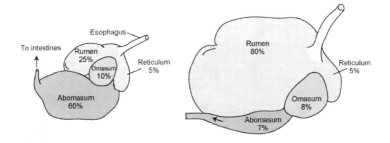

FIGURE 9.2. A schematic drawing of the development of the cow's four-chambered stomach. Only after a calf has begun feeding on grass does the rumen develop fully in size and function.

half of the rumen and that the cow's anatomy and physiology only become whole through the activities of feeding and digestion.

In the rumen, forage churns around in the fluid of the saliva and any water the cow has drunk. The rumen itself does not secrete digestive juices. When it is about half full, a wad of partially digested forage (what we call the cud) is regurgitated back into the mouth. If you are watching, you can see a bulge rapidly course up the cow's neck. When the cud reaches the mouth, the cow begins to ruminate. She grinds the food between her large cheek teeth in rhythmical, circling motions of the lower jaw. She chews a cud about 50 to 60 times before swallowing it. Soon thereafter another cud travels up the throat, and rumination continues. The saliva glands secrete copious amounts of saliva while the cow is feeding and ruminating—up to 50 gallons a day. Yes, that's right: 50 gallons.[1] The drier the feed (for example, hay), the more saliva a cow secretes, and the greater the amount of water she drinks.

Cows usually lie on the ground while ruminating, often with drooping or fully closed eyelids. If you are ever in a hectic state of mind and find yourself driving through the countryside and have the luck to spot a herd of cattle lying on the ground—I know, not too likely a scenario—stop and spend a half hour attending to the herd. Expand out into it. You'll calm down. As they lie quietly on the pasture, their activity focused inward on grinding and digestion, the cows radiate centeredness and quietude. For the total of eight or so hours of rumination per day, it is as though the mixing, breaking down, exchanging, and building up of

FIGURE 9.3. Dairy cows ruminating; Hawthorne Valley Farm, Ghent, NY.

substances is telling the cow an intricate and enchanting story that she is intently listening to.

As with bison and other ruminants, digestion in the rumen is facilitated by microorganisms that break down cellulose, the main, hard-to-digest component of fresh forage and hay. The forage is churned around, and it takes a few days for it to fragment into ever-smaller particles and to be broken down biochemically by the microorganisms. During this process nutritious fatty acids are released and absorbed through the rumen wall into the bloodstream. Since saliva is alkaline, it serves as a buffer and prevents the environment of the rumen from becoming too acidic. In an acidic environment, the microorganisms could not thrive.

Digestion is such a central part of the cow's life that even the animal's head plays a major role in breaking down the forage, through copious salivation and about 40,000 grinding motions a day in grazing. As biologist E. M. Kranich suggests, you can consider the cow's mouth functionally as a fifth chamber of the stomach.[2] After the mouth, digestion then continues in the microbial realm of the rumen. From there, the partially digested food moves into the other three chambers of the stomach, which continue the process of transformation. Only the last chamber (the abomasum) is comparable to our stomach. It secretes hydrochloric acid that kills bacteria, and digestive juices that break down proteins. As if the mouth and four stomach chambers had not done enough, digestion continues in the approximately 130-foot (40-meter)-long coils of the small intestine. (That's about 20 times the length of the animal!) After the cow has broken down the substances as far as possible and absorbed the many nutrients into the bloodstream, the large amounts of fluid that have been secreted as saliva and digestive juices are also reabsorbed, mainly in the last part of the digestive tract, the large intestine.

What has been digested and reabsorbed in the gastrointestinal tract enters the blood. The blood has the unique feature of being a fluid organ that connects all organs of the body by flowing through them. It gives over substances to the organs and receives substances from them. We need to imagine the blood as changing at every moment along its pathway. In every part of the body the blood is distinct inasmuch as it is responding to what comes from the organs and what they need. And yet in all this transformation, it remains a coherent flowing organ. Through this mediating activity of the blood, what the process of digestion brings forth allows the animal to continually recreate itself.

But that is not all. Through digestion, substances arise that the cow does not incorporate into her own organism, but rather gives off into the larger world. At the front end, she exhales with every breath—as all animals do—moist, warm air that is richer in carbon dioxide than the air she inhaled. Cows also burp frequently, in the process giving off methane-rich air that has arisen through ruminal fermentation. At the back end, she releases large amounts of urine and dung into the environment. A dairy cow weighing about 1,000 lbs. will excrete a total of 80 lbs. of urine and dung a day.

In contrast to the solid dung of other ruminants like sheep or goats, cows have fluid dung. The cow's large intestine does not absorb as much water out of what has been digested. We could say that from her moist snout, through all the secretions in her digestive tract, and finally in her dung, the cow embodies fluidity more than other ruminants—in a sense a paradox for such a large, heavy-boned and stout animal. The solid bones support a massive body and in the blood and the voluminous inner spaces of the digestive organs, continual and intense transformation occurs in the medium of fluids.

A most special fluid gift that the cow creates is milk. It provides just that nourishment her offspring need. And through domestication and husbandry, she creates more milk that we use for our consumption. Fill a glass with milk and place next to it a glass with grass in it. Two wholly different substances. The cow transforms the dry, fibrous grass into a nutritious creamy fluid. This demands intense activity on the part of the whole physiology of the cow. Breaking down and digesting grass already places high demands on the body. For example, for every quart of saliva the cow creates, 300 quarts of blood pass through the salivary glands. The other digestive organs are sustained by a similarly strong circulation.

The intense transformation of substances and the secretion of fluids characterizing the digestive process are heightened in the formation and secretion of milk. For every quart of milk, three to five hundred quarts of blood pass through the udder. The udder receives from the blood—and that means from the rest of the whole animal—the substances it needs for its mammary glands to create milk. Fine membranes separate blood and mammary glands. On the one side flows nutrient-rich blood, giving over proteins, water, fats, and carbohydrates to the mammary glands. And on the other side of the membranes the glands fashion and secrete a creamy white fluid. It is hard not to be in awe of the cow's ability to transform substances in its quiet and steady way.

For most modern consumers, milk is a packaged good that is found in the refrigerated section of a store. We probably have learned that this milk comes from cows, but many children who are growing up in an urban environment will never have seen a cow. We probably don't know what kind of dairy farm the milk came from or how the animals were fed and treated.

If, by circumstance or study, we do know something about these things, then we have begun, at least in our minds, to free the milk from its status as an isolated product for consumption. We can see it as an expression of a whole nexus of processes. The generation of milk stands as the result of the cow's interaction with her peripheral half—with pasture, soil, sun, weather, and of course with her human handlers. Which brings us to domestication.

FIGURE 9.4. The extinct aurochs (*Bos primigenius*); drawing based on a sixteenth-century painting. (https://commons.wikimedia.org/wiki/File:Ur-painting.jpg)

Domestication

There is much "darkness which shrouds the original achievement" of domestication in animals and plants.[3] In the Near East around 11,500 years ago, it appears that human beings were collecting and perhaps planting wild plants, and herding wild animals.[4] The wild, long-horned aurochs (*Bos primigenius*, also called urus), a large animal standing six feet high at the shoulders, lived in that region. It inhabited vast areas of Europe and Asia, and was hunted to extinction in the seventeenth century.

Despite its formidable size, people in ancient Mesopotamia evidently herded the aurochs. Over time, their physical, physiological, and behavioral characteristics diverged from those of the wild aurochs. They became domesticated. How this actually occurred is what remains shrouded in darkness, but by around 10,000 years ago in the Middle Euphrates and Tigris valleys of Mesopotamia, physically distinct domesticated cattle had arisen—and they were considerably smaller animals than their wild progenitors.[5]

Herding and domestication initiated a new and intimate connectedness of the life of human beings and animals. We changed the animals—

FIGURE 9.5. The Egyptian goddess Hathor depicted as a cow; from the Temple of Thutmosis II in Deir el-Bahary, Egypt. (Photo by Henry Edouard Naville, 1907; Wikimedia Commons)

FIGURE 9.6. King Thutmose III as boy suckling from the goddess Hathor, depicted as a cow. (Egyptian Museum, Cairo)

adapting them to our needs—and we in turn adapted to their needs. Domestication and the development of animal husbandry practices and breeding represent a coevolution of human being and animal.

What is all too easy to overlook is that the relation of the ancient peoples to animals always also encompassed a spiritual dimension, which I pointed to when discussing the Plains Indians and the bison. The ancient domesticators were not examining and breeding plants and animals with the utilitarian mind of a modern breeder. Nor were they half-rational tinkerers who somehow hit upon ways to domesticate animals in order to eke out a living (an all too common image of "primitive" peoples and "cavemen").

In the ancient Middle East, the cow and the bull were connected with goddesses and gods. The Egyptian goddess Hathor, for example, was often depicted as a cow or as a woman with horns resembling those of the aurochs. Horns were an especially revered part of the animal. They were a sign of vigor, power, beauty, health—and the divine in the world. Domesticated animals (and plants) were venerated because it seemed clear to the people of those cultures that the animals embodied the divine, and the divine worked in and through them. Animal sacrifices to the gods and the gift of milk to the gods need to be seen in this light—they strengthened the connection to the divine. Domestication was one feature of the effort to intensify the union with the divine. It was not motivated simply by economic concerns.[6]

During the thousands of years that traditional pastoral and farming cultures coevolved with animals and plants, the demanding day-to-day interactions did not stand alone as "a job." The work was surely hard, but it was infused with reverence and enhanced by rituals and periodic festivals to give thanks and to celebrate life. The life and work with animals was embedded in a larger cultural and spiritual context that gave rich meaning to both the human deeds and the animal beings.

In the course of domestication of cattle, a large number of breeds arose around the world. Currently you can find lists naming 800 cattle breeds, about 220 of which are listed as dairy cattle.[7] Many of the older, traditional breeds were raised for meat and dairy, and may have also served as draft animals. The breeds differ in size, shape, temperament, and many other features. This wealth of variety—which was not present in the wild progenitors—has been brought to appearance by the ongoing and intimate human-animal interaction. It is therefore not surprising that Charles Darwin could draw on domestication as an exemplar for evolutionary change—for the development of new varieties of animals and plants.

In this evolutionary process, the domestic plants and animals have become dependent upon human beings. We have brought them into the circle of our lives. In the process they lose some characteristics of their wild relatives and gain new ones that strengthen the connection with humans.[8] Domestic animals are open to human beings interacting with them; they do not flee. They are more submissive and amenable to living in corrals and confined quarters. They also lose some of their vigilance in noticing and avoiding predators. Such characteristics are also seen in young wild animals, so some of the changes that have occurred through domestication can be viewed as the retention of juvenile characteristics in adult animals. Such retention can also be found in the body, for example in the relatively short jaw or in the greater deposition of fat under the skin and in the muscles.

Domestication has led to a number of childlike (paedomorphic) characteristics in the animals so that they are open to and in need of day-to-day tending. But in contrast to a human parent-child relation, with domestic animals the responsibility of close, day-to-day tending never ceases throughout the whole of the animal's life. There's not a point at which we can say to a cow, "Okay, go out and see what you can do on your own." Domesticated animals that do escape and survive (so-called feral animals) usually develop—as we could expect—in behavior, physiology, and even in morphology, characteristics resembling those of wild animals. If they survive, they re-wild.

Each domestic animal has evolved characteristics that we desire and make use of, be they milk, meat, wool, or eggs. Inasmuch as breeders focus on achieving specific characteristics (leaner meat, more milk, etc.), the animal and its life become increasingly one-sided, as the recent history of dairy farming shows.

Ever More Milk

Until the twentieth century a cow gave about as much milk per day as her calf would have drunk—about a gallon per day in present-day breeds.[9] In 2018, the milk production of high-milk-producing dairy cattle—mostly the Holstein breed—reached in the United States an average of nearly nine gallons per day.[10] This increase has taken place essentially since World War II.

A high-producing dairy cow today gives about five times as much milk as a cow at the end of World War II. And today, there are only about a third as many cows giving all the milk.[11] How has the remarkable increase been achieved? At the source lie the goals of achieving greater efficiency and productivity. These are ideas—motives—that come from human beings. The cow has become the medium for their realization.

There has been a single-minded focus on breeding and selecting cattle that produce ever-more milk. Which cows are the highest producers? Which bulls have offspring that produce more milk? A majority of dairy cows are artificially inseminated with the sperm of select bulls that have had the highest producing offspring.[12] Semen is sent around the country and the globe. Most dairy cows today are Holsteins (86 percent of all dairy cows in the United States).[13] They stem from a relatively small group of bulls. For example, one Holstein bull born in 1962 (named Chief) had a daughter that produced an unusually large volume of milk. Chief became a sought-after bull for mating, and in the end he fathered "16,000 daughters, 500,000 granddaughters, and more than 2 million great-granddaughters." His sons also became favored sires. As a result, the lineage of 14 percent of Holstein dairy cattle in the U.S. can be traced to Chief.[14]

In part, the increase in production has been achieved by selecting animals that were larger and therefore gave more milk. Holsteins are the largest dairy cows and the highest milk producers. Also, instead of milking twice a day, over half of the large dairy farms (with more than 500 cows) milk cows three times per day.[15]

Of course, if a cow grows larger, it needs to eat more to sustain itself and to produce more milk. Therefore, much effort has been put into finding a combination of feed that supports greater—more efficient, as one likes to say—milk production. Instead of being fed only fresh pasture forage or hay, especially high producing dairy cows are also fed concentrates, which can make up half or more of their food.[16] These concentrates are mixtures of grains (most often corn), soybeans, and other plant-based substances that are rich in protein, starch, fats, vitamins, and minerals and allow a cow to produce more milk than she otherwise could. One nutrient supplement that is fed to about a third of cows on large farms is blood meal from slaughtered animals—a source of protein and minerals, but not exactly what you would think a herbivore would naturally eat.[17]

Additionally, various humanmade "supplements" have been used to increase growth and milk production. For example, in the 1940s it was discovered that antibiotics not only kill bacteria in the gut but also increase growth in chickens and livestock.[18] So dairy farmers began including antibiotics in feed or water to stimulate growth. In 2013, over three-quarters of large dairy operations with more than 500 cows administered antibiotics prophylactically in feed or water to weaned or pregnant heifers (to promote growth or prevent disease).[19] Another, more recent, "supplement" is genetically engineered bovine growth hormone (so-called rBGH). Since the 1990s many conventional farmers inject their dairy cows with it to stimulate increased milk production.[20]

While all the breeding, feeding, and technological innovations (such as milking machines and milking parlors) have directly influenced increased milk production, it would likely have never gone so far had there not been support through government subsidies. Since the 1940s in the United States, overall consumption of milk products has risen with population growth, but in recent decades production each year has far surpassed the demand.[21]

This has been possible because since 1949 the federal government has been supporting the overproduction of dairy products through its milk-price-support program. From that time until the turn of this century the USDA "stood ready to buy as much butter, nonfat dry milk, and Cheddar cheese as manufacturers wanted to sell at specified support purchase prices."[22] While this automatic support system no longer exists in that form, it played a major role in encouraging overproduction for many decades. Today the government continues to provide subsidies to the dairy industry, and overproduction continues. In 2016 the government spent $20 million to buy up 11 million pounds of surplus cheese.[23] And as of May 2017:

> The U.S. has more than 800 million pounds of American cheese in reserve, the most since 1984, according to the USDA. The amount of butter in reserve totals 272 million pounds, the most since 1994. Some U.S. farmers are dumping millions of pounds of excess milk onto fields. In the Midwest and Northeast, nearly 78 million gallons of milk have been dumped so far this year, up 86% from the same period last year.[24]

Government subsidies directly support not only dairy farmers, but also farmers who grow crops such as corn and soybeans that are fed to cows. In the years between 1965 and 1990, for example, overall yearly payments by the government to farmers of all kinds averaged $10 billion per year.[25] When we buy inexpensive food, we need to realize that the price at the store does not include what we pay through our taxes.

The Larger Context

If you are a proponent of higher production, efficiency, specialization, intensification, and growth, then the dairy industry since the middle of the twentieth century is exemplary. The increase in milk production *is* a remarkable achievement when looked at in isolation—in isolation from the larger environmental, cultural and economic contexts, and in isolation from the fact that a cow is a living being and not a production machine.

The increase in milk production in the second half of the twentieth century was an integral feature of the industrialization and mechanization of agriculture that was supported by the government as I described above. Fewer cows produced more milk, and those cows were kept on ever bigger, highly mechanized dairy operations. According to USDA statistics, the historical maximum number of dairy cows in the United States was reached in 1940—24.1 million cows. These cows were part of 4.6 million small family farms, virtually all of which had 30 or fewer cows.[26]

These were not "dairy farms" in the modern sense of a single-product farm. Most had some beef cattle, perhaps pigs and chickens, and they grew crops to feed their animals and themselves. On these family farms, cows usually had access to pasture during the growing season and were fed hay and silage during winter. Each cow was known to the farmer and was milked for many years.

All this changed radically in the coming decades. By the twenty-first century there were far fewer farms—in 2012, for example, only 64,000 dairy farms housed the 9.3 million dairy cows. Sixty percent of the animals were on only five percent of those farms, namely those with more than 500 cows.[27] These large farms focus only on milk production and have huge barns housing the cows. Operations with over 2,000 milking cows are no exception today, and each year they get bigger. On these farms, the cows are rarely or never set out on pasture to feed and freely move around. About 80 percent of all dairy cows in the U.S. today have no access to pasture while they are giving milk, which is most of the year.[28] They remain in barns or loafing sheds with little movement and no interaction with their peripheral half—the pasture.

It is perhaps not surprising that one-sided breeding for higher milk production, the provision of grain-rich feeds, the confinement of many animals in close quarters, the guiding mind-set that views animals as production units, along with an economic model that aims at efficiency, concentration, and more output, all set the stage for an array of problems for cows.

It has long been known that breeding and feeding for high milk production in cows living in confinement conditions affects the health and
vitality of the animals.[29] The turnover of cows in a large dairy farm is
significant—a fifth of the cows may be killed after their first lactation due
to fertility issues, and, overall, such an operation may kill up to 40 percent
of its cows every year. We should let this fact sink in. A typical Holstein
today has two to three lactations in her life before she is killed—for a
variety of reasons (less productivity, lameness, mastitis or other diseases).
So an average high-producing Holstein will live only four to five years,
while the life span of cattle in general is more like 20. (Just to note: beef
cattle live even shorter lives—one to two years.) Without the demand to
produce as much milk as possible in a short period of time, a cow will
reach its peak of milk production after three or four years of lactation and
can continue healthy lactation for a number of years beyond that.

When you feed specialized animals food that deviates significantly from
their natural diet, you challenge their digestive system and in turn their
whole organism. When you feed cows high-grain concentrates, you feed
them less forage, meaning less roughage and fiber. As a result, mixing
motions in the rumen, burping, rumination and saliva flow all decrease.
Acids resulting from the microbial fermentation in the rumen increase,
leading to a more acidic environment.[30] This is called subacute ruminal
acidosis and can have grave consequences, such as "feed intake depression,
reduced fiber digestion, milk fat depression, diarrhea, laminitis [lame-
ness], liver abscesses, increased production of bacterial endotoxin and inflammation characterized by increases in acute phase proteins."[31]

The inflammation of the udder—mastitis—is another common problem among high-producing cows.[32] Since it is an infectious disease, strict
hygienic procedures help prevent bacteria from entering the udder via
the openings in the teats. But this is only one side of the problem. Due
to the intense circulation in the udder during lactation, the udder is sus-
ceptible to inflammation. (Increased circulation always occurs in inflamed
organs—it calls forth the warmth and redness of inflamed tissue.) When
milk production is increased to the utmost degree, the udder is almost on
the verge of inflammation even without bacteria. The cow's physiology is
stressed, her immune system taxed, and when bacteria do enter the udder,
mastitis is likely.[33]

As I mentioned above, antibiotics are routinely added to feed and water of dairy cows before they start giving milk. They have a twofold function—to increase growth and to prophylactically protect against infectious diseases such as mastitis. These applications have led to an enormous use of antibiotics in dairy cattle (and also in pigs and chickens). About 80 percent of all antibiotics used in the United States are administered to livestock and chickens.[34]

In these non-clinical uses of antibiotics, smaller amounts are used than in acute cases; as a result more bacteria survive and some become resistant to the antibiotics. Over the past decades bacteria have become resistant to virtually every antibiotic, causing a global crisis—not only for the treatment of animals with acute bacterial infections, but also for human beings, since therapeutic antibiotics for humans have also been used for animals, and resistance can be transferred from one type of bacterium to another. In the case of dairy cows, the resistant bacteria can spread in a variety of ways. Most resistant bacteria are found in manure, since only a portion of antibiotics is taken up by the cow's body, and the rest is excreted. The manure can find its way into streams and groundwater, contaminating them. Bacteria can also spread through milk—although, generally, pasteurization will kill most bacteria. Workers at dairy farms who handle the cows and manure can carry the bacteria out of the farm, just as can the meat of cows that have been slaughtered.[35]

The increased use of antibiotics to treat or prevent ailments became "necessary" because the animals have been kept in conditions that are decidedly unhealthy for them. Instead of changing those conditions, a technology is applied that itself leads to more problems, which then ripple out from the unhealthy centers of concentration into the broader environment. There is no way to keep the effects of the extreme conditions on a large dairy farm from impacting both the being of the cow and the larger world.

Horns and Tails

If you look into a large open barn housing dairy cows, you can notice that they have no horns and may well have only the upper part of their tails. This is not because they were born that way. It is because as calves the developing buds of the horns were killed, most likely with a searing iron, and three-quarters of the tail was cut off.

This "disbudding" of dairy cows is so widespread today that many people don't realize that they are looking at animals whose ability to grow a part of their body has been truncated. Hornless cows have become the norm. And, in fact, in the province of British Columbia a legislative act encourages the dehorning of cattle (dairy and beef) and penalizes farmers or ranchers who raise horned cattle.[36] Tail docking, as it is called, became more widespread as dairy farms got bigger and more animals were confined in smaller spaces.

Both disbudding and tail docking cause the animal pain and are not trivial interventions—they are not comparable to clipping nails.[37] Although veterinarians recommend local anesthesia and analgesics to reduce pain, they are in fact rarely administered by American farmers.

So why do modern farmers want cows to be hornless and tailless? What would motivate them to submit their animals to painful procedures that result in them lacking two organs that they by nature possess and that are normally integral to the animal's life and behavior?

Horns are seen as dangerous. When you read about mainstream dairy farming, you will find statements such as this one in the *Journal of Dairy Science:*

> Handling and management of horned animals is deemed
> impractical for human and animal safety. Horned dairy cows
> pose a risk for stockpersons during routine management practices
> (milking, hoof trimming, calving) and veterinary examinations.
> Moreover, horned animals can cause injury to herdmates during
> aggressive interactions and competition at the feeding gate.[38]

What statements like these don't tend to mention is that the main problem arises because the dairy industry has chosen to house cows in close confines. In the case of free stalls and loafing barns, cows move around, but because they are in a relatively small space, they do not have the freedom of movement they have on the pasture, where animals can easily give space to one another and retreat if needed. When hemmed in, cows get agitated and can be more aggressive, so it is not surprising that under such conditions cows with horns may hurt each other and their handlers.

The primary response to this human-created problem is to disbud the cows. As the authors of a review article entitled "To be or not to be horned—consequences in cattle" conclude, "Disbudding or dehorning are measures to adjust animals to husbandry conditions that are insufficiently adapted to the species-specific needs of cattle."[39]

And why tail docking? Proponents say it helps the cow stay cleaner and supports good hygiene, but none of the scientific studies that have been carried out support that view.[40] In the end it comes down to increasing the comfort of the handlers—they are no longer bothered by swishing tails. Once the tail has been amputated, a cow can no longer bat flies from her hindquarters. For a cow this means that more flies gather on her hind end and she has no means of removing them.[41]

Both horn disbudding and tail docking are surgical procedures through which farmers adapt cows to conditions humans have created, conditions that they feel warrant such interventions. They did not ask the cows, "How might we interact with you, so that you can live out your life in a species-appropriate way and at the same time you can serve us?"

This question reveals the central quandary that we can't avoid if we want to interact with domestic animals in a conscious and responsible way. I don't think there are any easy answers, any recipes or prescriptions for one right way to act or not act. But I do think we can make every effort to address the question and the thorny issues it entangles us in. It is all too easy to sleepwalk through our interventions with domesticated animals.

We can begin by asking: Are horns and tails expendable organs? From a purely human-centric, utilitarian point of view, the answer is clearly yes. Cows can live without them and still produce milk. But what about the cow's point of view? This is where it gets hard. We can't ask the cow directly, but we can study the cow and try to understand how horns and tails are members of its whole being.

From an evolutionary perspective, tails and horns are integral members of the hooved mammals, to whom cattle belong. As I mentioned in earlier chapters, all horn-bearing mammals are ruminants with four-chambered stomachs, have paired hooves, and possess no incisors or canines in their top jaws. These—with other features—represent the particular coherence of this group of animals. You find this suite of characteristics in countless variations extending back into the deep history of mammalian evolution. The inner connection among hooves, horns, a four-chambered stomach, and the absent upper canines and incisors may not be transparent to us, but the fact of their long evolutionary codevelopment shows that they do have a connection even if we can't fathom it.

So horns are not "add-ons" to this large group of animals.[42] When paleontologists find fossils, the discovery of the bony cores of horns that emerge from the frontal bones of the skull is key in determining whether the animal is part of the bovid family, which includes, besides cattle, such animals as bison, Asian and African buffalo, wild sheep and goats, and the large variety of antelopes. The to-date oldest known bovid lived in the early Miocene period, about 18 million years ago.[43] And tails have been around even longer.

In the first months of its life, a calf begins to form the two horn buds. Initially these are not connected with the skull itself, but are free-floating in the skin above the skull. The buds then attach to the frontal bones of the skull. The horns have a bony core covered by what we see outwardly as the horn—a dense protein sheath called keratin (which is also the substance that makes up hair, hooves, and nails in mammals). The bony core does not remain solid; it becomes air-filled, and its inner surface is covered by a mucous membrane. The horn cavities are extensions of the large frontal sinuses, which connect through small passages with the nasal cavity.[44] Because these passages are narrow, air is exchanged between the nose and sinuses only slowly. Nonetheless, it is fascinating to realize that air penetrates all the way into the very core of the horns and that the air in the upper skull and horns communicates with the air that a cow breathes.

When calves are disbudded, the skull develops a different shape. The area at the back top of the skull—which would have been between the horns—rises up in the middle and becomes more pointed. Concomitantly, the skull does not grow as wide, showing that the horns play a role in the overall shaping of the skull during development.[45]

The horns are amply supplied with nerves and blood vessels and grow throughout the cow's life. Growth rings on the base of the horn sheath indicate how many lactations a cow has had. Although the outer sheath is not itself living tissue, the cow is very well aware of her horns and the extent of their reach. For example, a cow in a stall can access narrow feeding racks by tilting her head to come between the bars without touching them. And farmers report that a cow can with the tip of a horn deliberately open a closed feeding rack.[46] With her horns a cow scratches herself, and she can spar with other cows' horns. Horned cattle tend to keep a larger space between individuals, than do ones that have been dehorned, which may relate to a sense of larger body space that the horns mediate.[47]

So the horns are clearly not a meaningless appendage, nor is the tail. The tail is the extension of the vertebral column and consists of a series of ever-smaller vertebrae, muscle, nerves, and skin that at the tip grows long hairs. With this highly mobile organ the cow can effectively swat flies and pests that land on the different parts of her hindquarters. It is also an expressive organ, although little do we fathom the variety of inner states the tail's movements may be giving voice to.

Diminishing the Animal

We can recognize that organs such as the horns and tail are integral to the life of the cow. But it is also a fact that the animals can live without horns or a tail. Since—as one might argue— we have already changed the animals considerably through breeding and animal husbandry practices, why not dock tails and remove horns, especially if we take measures to reduce pain and discomfort?

It is quite easy to follow the logic of this argument. It is inherent in many efforts to continue to alter animals in even more radical ways. Take, for example, an article in the journal *Neuroethics* entitled "Knocking Out Pain in Livestock: Can Technology Succeed Where Morality has Stalled?"[48] The author notes that factory-farmed livestock suffer from a variety of ailments that can cause pain. He assumes that with the increasing consumption of meat on the planet, factory farms will continue to be necessary. He does not consider that there are alternatives to factory farming. So, in his argument, if animals will continue to suffer under the factory farming conditions we have created, and we decide not to change those conditions, then the solution is to genetically engineer livestock so that they do not sense pain. There are indications that something like this works in mice, so maybe it would work in livestock. He presents this solution as a good way of addressing a critical animal-welfare issue.

The aim is to "help" animals by making them less-than-animals, by diminishing their animalness. This is a grotesque view. You recognize that you have created conditions that do not allow animals to flourish—that cause them pain and suffering and that alter fundamental characteristics— and then, instead of saying, "It's high time to change those conditions," you say to the animals, "Let's diminish your capacities so that you don't notice how bad off you are." Or as philosopher Marcus Schultz-Bergin puts it, in such a perspective you are blaming the victim, and "You could imagine us telling the animals 'if you were just not capable of suffering, then we would not have to make you suffer.' This seems quite perverse."[49]

Much of modern breeding and husbandry in service of industrial agriculture does in fact lead to diminishment and suffering on the part of animals. How could it be otherwise when we view and treat animals as commodities (things to be bought and sold), resources (things to be mined), and even as bioreactors (living factories to produce novel substances)? Since instrumental, utilitarian consciousness dominates the way animals are kept, bred, and treated, the animals themselves can have no voice. Once you have instrumentalized animals in your mind, then you can always find a justification for further manipulations that suit your agenda.

And once you have gotten used to viewing and treating animals in a particular way, you no longer realize what you are doing. When researchers surveyed 113 dairy farmers about their operations, the farmers usually stated that "they were treating their cows well, because they follow the recommendations of university and veterinary specialists."[50] But the researchers noticed that when the farmers spoke about the quality of the cow's life, they "seldom mentioned a cow preferring pasture." Their practices had become the norm, and they didn't have in mind what is integral to the species. As the researchers remarked, the farmers "often are not considering the view from the perspective of a cow."

Do domestication, breeding, and husbandry practices necessarily lead to "diminished" animals? Many years ago, I had a conversation with an environmental ethicist who was concerned about humanity perceiving and articulating the intrinsic value of other-than-human nature. Interestingly, he considered domesticated animals to be, in a sense, degenerate, since they are no longer capable of living in the wild. In this

Figure 9.7. Part of Hawthorne Valley Farm's dairy herd. Note the bull and a couple of calves in the foreground. A herd is only complete with cows, calves, and a bull.

view, breeding of animals leads in due course to their diminishment. While I can understand his perspective, I also could describe to him the childlike (paedomorphic) characteristics of domesticated animals I mentioned above, which in my view cannot be considered diminishment. Such characteristics form the basis of the close and evolving relation between human beings and domesticated animals. Through thousands of years these animals have become an integral part of our lives and we of theirs. Increasingly, we have dominated the interaction, while our dependence on domesticated animals has never ceased. The question is, can the weights on the balance shift? Can we give the animals a greater voice in our dealings with them?

Dairy Cows at Hawthorne Valley Farm

I live and work close to the 900-acre biodynamic Hawthorne Valley Farm in upstate New York that milks around 65 to 70 dairy cows. I know the farm quite well. My wife helped start its Community Supported Agriculture program (CSA), and we have been members ever since. I have been involved in the education program for the farm apprentices for more than a decade. And when my children were young, they spent many mornings in the barn helping with the cows.

This is no typical, single-purpose dairy operation. It has many different facets. The farm has its own dairy and the cows' milk is sold at the farm's store as raw milk, and is also made into cheese and yogurt that are marketed more broadly. The farm raises pigs and chickens. The CSA garden provides about 300 families with vegetables during the growing season and also offers families storage vegetables in the winter. A market garden sells vegetables at Green Markets in New York City. A farm store sells the farm's products, but is mainly a full-line natural food store selling mostly organic and many local products. The farm has a learning center that serves hundreds of children each year; some come with their school classes and spend a week on the farm, others come to a variety of summer camps. Each year five to seven apprentices work and learn on the farm.

The farm operates under a nonprofit umbrella organization (the Hawthorne Valley Association) of which a number of other initiatives are a part. For example, a Farmscape Ecology Program seeks to "foster informed, active compassion for the ecological and cultural landscape of Columbia County, NY through participatory research and outreach;" and the Hawthorne Valley School is a K–12 Waldorf school located across the road from the farm buildings. (I taught biology at this school for nine years starting in the early 1990s.)

When you enter the valley, you often pass by the cow herd grazing or ruminating in a pasture. All the cows have horns. It is a mixed herd— mainly Brown Swiss with Jersey, Guernsey, and Ayrshire influences. A bull is always with the herd, and young calves roam playfully, nurse, and make their first attempts to graze. For about half the year (May through October), the cows are on pasture all day and night except when they are being milked. During the winter and early spring, when the grass is not growing and the temperatures can be very low, the cows stay mainly in a large free-stall but always have access to the outdoors. They are fed hay and silage from the farm. They are not fed concentrates.

I spoke recently to the former longtime farm manager, Steffen Schneider, and current farm manager, Spencer Fenniman, and asked them about their untypical husbandry practices. What motivates them? What are their challenges?

At the heart of biodynamic agriculture lies the intent to view and work with the farm as a kind of differentiated organism in which the manifold activities and beings support and enhance each other. The farmers strive to recognize the unique contributions of each of these aspects—such as the weather, topography, soil, compost, plants, animals, and human beings—and want to facilitate their interweaving to create a healthy farm and healthy products. The nonprofit Cornucopia Institute gave Hawthorne Valley Farm its highest rating ("beyond organic") for diverse, small-to-midsized dairy farms that emphasize pasture and forage-based feed.[51]

In respect to their dairy cows, Schneider and Fenniman emphasized that they would like to allow their cows as far as possible a free range of expression. That includes living in a herd, having horns, and being on pasture to the extent the climate allows. All this contributes to the health and well-being of the animals. For these farmers, the herd is especially important. If a new bull is brought into the herd, it joins as a calf, so that from a young age it is part of herd life. The bull mates with the cows on the pasture at will; there is no artificial insemination. When cows give birth to female calves, these calves stay with the herd and are raised to be dairy cows. In this way, family lines are maintained for many generations. The herd is integrated into the land and the husbandry practices, and it develops as a kind of extended organism through time.

As the farmers emphasized, each cow has its own temperament and unique characteristics. This makes the herd a place of dynamic interactions between individual cows of different age groups and dispositions. Sometimes a cow or bull that is too aggressive and doesn't meld with the rest of the herd is culled. In the social dynamics, each cow has its own social space, and the horns and the way she uses them are part of that social interweaving. I asked if there were many injuries—to cows or to farmers—because of the horns. The answer was, "Only when we make mistakes." In other words, as long as you know how cows use their horns and give them adequate conditions, injuries are rare.

A cow will be milked as long as she remains healthy and still produces adequate amounts of milk. Recently, a 19-year-old cow was retired that had been milked for 16 years! A vet described her a couple of years before as having "legs of a five-year-old cow." Now, this cow was an exception, but often cows at the farm will be milked until they are 10 years old (about seven lactations). Around this age, the milk production often drops, but if it doesn't, a cow may be milked for some years more. In this way the farm gives the cows the possibility of living longer lives within the herd, which in turn affects the herd dynamics, since all age groups of cows are represented.

The farm does not breed cows for maximum milk production. Of course, the cows need to produce a fair amount of milk—milk that is well suited for drinking, but also for processing into yogurt and cheese. But the farmers consider even temperament and adaptability to be very important. The cows should thrive well in the herd and be amenable to diverse interactions with human beings—diverse milkers and handlers, and children who come and go. Recently the farm has begun introducing, via bulls, the Normande breed and Milking Shorthorns into the herd. The goal is to have a more multipurpose herd that produces both milk and good meat. While the farm presently sells meat from culled cows, they hope to improve the quality of the beef. The farm is currently raising some of the bull calves, which were previously always sold soon after birth.

Given the fact that the cows live a fuller cow life and are not pushed in only one direction, it is not surprising that they do not produce nearly as much milk per lactation as does a high-producing Holstein in a factory farm operation. While a Holstein that is milked two to three times a day may produce eight to nine gallons per day, a Hawthorne Valley cow milked twice per day gives around four gallons per day. A Holstein cow in a factory farm has on average only two to three lactations in her lifetime, giving large amounts of milk in her short life. In contrast, the Hawthorne Valley cows may have between six and ten lactations in their longer lives, and the total amount of milk they give in a lifetime may be less, but approaches that of the short-lived Holstein. Through this practice, the cows' physiology is much less stressed, and the farm has lower costs by raising animals that live longer.

The farmers see milk as a gift, and without the milk the farm could not survive economically. Surprisingly—at first hearing anyway—Schneider emphasizes that "in a holistic picture, in many senses, the manure is the primary gift that the cows give us." Manure is a key element in building soil fertility. On the one hand, cows leave urine and feces on the pasture, and, on the other hand, manure collected in the barns is mixed with straw, other plant matter, and food scraps from the store and school, then composted, and finally spread on the fields. On factory farms, manure has become a waste issue—the main task is often getting rid of it. On this farm, manure is a precious gift to be transformed and given back to build fertile soil.

Vegetable growth demands more from the soil than does pasture. As a result, the farm spreads about one-third of its compost on the vegetable fields, which make up only 2 percent of the agricultural land. As Schneider commented, "For every head of broccoli, you have to imagine a cow in the background…. I've not seen a study that shows you can practice sustainable farming without livestock to build soil fertility." So in this sense, the dairy cows are integral to growing vegetables for human consumption. And by utilizing manure and plant matter that come from the farm to create compost, the farm does not rely on nonrenewable fossil fuel products—artificial fertilizers—to promote plant growth.

Most small dairy farms are struggling, and many are closing. The price farmers receive for milk dropped markedly from 2014 to 2015 and has remained low ever since. Meanwhile, other costs continue to rise, so it is increasingly difficult to pay bills, and many farmers have large debt loads. The number of small dairy farms has been declining for decades. For example, in 2003 there were 70,000 in the United States; in 2017 there were 40,000. You only have to travel to the Midwest, where smaller farms of all kinds continue to disappear, to witness the demise of farming-based, small town culture.

Income from milk can often be less than it takes to produce the milk. To keep the farms going—and thanks to automation and to the cheap labor (often performed by undocumented workers)—farmers are commonly dependent on off-farm income earned by taking other jobs. This is the case in farming in general, with the exception of the largest commercial farms. Off-farm income can make up half of the income of smaller farms.[52]

The dire situation is revealed in the increase in suicides among dairy farmers.[53] The Center for Disease Control found in a 2016 study that of all occupational categories, farming, fishing, and forestry have the highest rate of suicides—3.5 deaths per 1,000 individuals.[54]

At Hawthorne Valley Farm, you have a different picture. It started as a small initiative in 1973 and has grown steadily since then. With its variety of farming, commercial, and educational activities, it has attracted more and more people to this rural area of upstate New York. It is one example of numerous initiatives in the Hudson River Valley corridor that are helping to establish a regional foodshed. But what is clear is that the farms cannot exist on their own. They need communities, and they need to be integrated into innovative economic and cultural contexts.

Reflecting on especially the farm's dairy operation, Schneider said that "being embedded in what I call a micro food value chain is the only reason we are still in business." What he means is that the farm sells its milk—which is certified organic and biodynamic—to its own dairy at a premium price, and the farm's dairy adds value by turning the milk into yogurt and cheese. The raw milk is sold at the farm's store, as are yogurt and cheese. The latter are also sold at New York City Green Markets, and the yogurt finds its way into stores in many eastern states. By adding value and services, the farm receives significantly more income than it would if it sold its milk on the bulk-milk market.[55]

On this farm it is clear that the farmers strive to pay attention to the characteristics of the cow, recognize the cow as a living being, and work with the cow's needs and nature in designing the dairy farm. But you cannot do justice to the cows in isolation. The land, plants, other animals, and the human beings who visit and work on the farm all need to be taken into account. This entails the significant and yet inspiring challenge of working within the larger ecological, social, and economic contexts in innovative ways.

Taking Responsibility

It is both our gift and our burden as human beings that we can become conscious of the larger reality that we are a part of and that we affect through our actions or inactions. It is in this awareness that the feeling of responsibility and the desire to take responsibility can arise.

When we interact with cows—or any other part of the world—our actions leave an imprint. Because cows are domesticated animals, they bear the effects of human interactions that reach back for thousands of years. We are responsible for these interactions and their effects. When we speak of *taking* responsibility, we are pointing to a conscious undertaking, not something that just happens through tradition or directives from authorities. We know ourselves to be in a relationship, and we know that it is possible to act in better or worse ways. We may not know what to do; we may not know what it would mean to act truly responsibly, but we are not sleepwalking through the day as if our thoughts and actions did not matter.

Taking responsibility is no easy matter. As we have seen, the conventional farmers who stated that their cows are, from an animal welfare perspective, better off than they were two decades ago, are judging the well-being of their cows within the confines of the existing system of industrial agriculture. They cited, for example, improved nutrition (concentrate feed and additives), better veterinary care, better ventilation in barns, and free stalls in which the animals can move around. I give these farmers the benefit of the doubt that they believe this. What I don't know are the motives that led them to make these changes—whether they came out of concern for the cows, concern for profits, or pressure from outside groups, to name a few possibilities.

But it is striking what the interviewed farmers did not mention. They did not state that the cows had been dehorned and their tails docked, that they were typically slaughtered after only a few lactations due to the stresses of high milk production and a variety of ailments, and that they generally had no access to pasture.

The farmers viewed the cow's welfare from the perspective of improvements within the already existing industrial management system. They were thinking within that box. They had lost sight of what was not in that box—the cow as a horned, tailed, and grass-grazing animal. When they say that the cow has better nutrition today, they mean in terms of what the cow needs in order to survive (I won't say thrive) within the industrial model and continue to produce ever more milk.

When a cow is considered primarily in terms of production, its reality as a living being recedes into the background. There is then no need to be particularly concerned about removing horns or tails, if that serves the larger goals of management efficiency. And the idea of genetically engineering cows to feel no pain may call forth no scruples. This notion can seem a consistent and logical extension of the trajectory that has been followed for decades. What is the problem, proponents of such approaches may ask? How easy it is for the mind to become caught up in a particular worldview that provides the frame for what is deemed acceptable.

Conventional farmers are themselves entangled in a system they often feel they can't escape. They may feel they have no choices other than to continue within the status quo or to abandon farming altogether. These are, I believe, not their only alternatives. The example of Hawthorne Valley Farm shows one. But I also think that most of us know situations in life in which we feel caught and can see only a couple of bad alternatives.

So how is it possible to break out of a system, and what facilitates and motivates moving beyond a worldview that promotes that system? Of one thing I'm certain: there is no one answer to these questions, and there are no recipes.[56] Keeping that in mind, I'd like to describe one example of the process of breaking out of a system and a worldview into a broader, more encompassing outlook.

The 22-year-old Aldo Leopold was leading, in 1909, a crew for the newly formed United States Forest Service. The crew was carrying out an inventory of the locations, quantity, and quality of timber in Arizona and New Mexico. It was wild country, and there were still many wolves. As Leopold later wrote, "In those days we had never heard of passing up a chance to kill a wolf.... I thought that because fewer wolves meant more deer, that no wolves would mean hunters' paradise."[57] Wolves were widely considered vermin—pests to be gotten rid of. Leopold wholeheartedly accepted this view. In addition, he had hunted since an early age and was still, as he put it, "full of trigger-itch."[58]

When he and his crew noticed, from up on a rimrock, an older wolf and her pups emerging from a turbulent river below, they immediately began shooting. They then climbed down to the banks of the river and found the old wolf lying on the ground, still alive but unable to move. Before their eyes, the wolf died. More than 30 years later Leopold wrote, "We reached the old wolf in time to watch a fierce green fire dying in her eyes. I realized then, and have known ever since, that there was something new to me in those eyes—something known only to her and to the mountain."[59]

As he watched the light in the wolf's eyes disappear, Leopold met the wolf for the first time. For a split second he glimpsed the wolf as a being in its own right. The impression stayed with him. In a sense, the wolf became part of Aldo Leopold on that day. And yet, it took a long time for the wolf to become a force in his thinking. The perception of the dying wolf was not enough. He needed to spend many years out on the land, where he observed the increasing effects of wolf eradication on the larger environment. "I have watched the face of many a newly wolfless mountain, and seen the south-facing slopes wrinkle with a maze of new deer trails. I have seen every edible bush and seedling browsed ... to death."[60] He also witnessed the effects of overgrazing cattle and sheep. As he remarked, "While a buck pulled down by wolves can be replaced in two or three years, a range pulled down by too many deer may fail of replacement in as many decades."[61]

First Leopold was jolted by the experience of the dying wolf, and some-thing opened up. Then he spent decades observing the manifold effects of wolf eradication. Gradually his worldview transformed. He broke through the boundaries of the notion that fewer wolves meant more deer meant great hunting. In the last decades of his life he worked to develop an ecological view of wildlife and worked tirelessly to protect wildlands. Toward the end of his life he formulated what he called a land ethic that "changes the role of *Homo sapiens* from conqueror of the land-commu-nity to plain member and citizen of it. It implies respect for his fellow-members, and also respect for the community as such."[62] His worldview had totally shifted—and the world that he perceived and incorporated into his thinking was much larger and more encompassing than the one he knew as a young man.

As it was with Leopold, it is not uncommon today that we wake up to our responsibility when we experience how human action has wrought havoc in the world. How often do we need to experience destruction—the killing of the wolf, the destruction of ecosystems—of life, to realize the value of the life that has disappeared? We feel: that's not right and something needs to happen. This is an important realization, but it is not enough. We need to find understanding and ways of acting that bring healing.

It is clear that when we view cows as production units or wolves as vermin, we are considering them both in far too narrow terms and primarily from the perspective of our own gain. We have avoided considering much of the reality of the animal's life and the way it is woven into the larger world. So we can begin to take that larger fabric of life seriously and turn our attention toward it. We can learn how the cow (or the wolf) is a truly integrated organism with a very specific way of being. We can realize that the cow is a giver of gifts, as farmer Steffen Schneider put it—the gifts of milk and manure—and is a substantial presence from which the farmers and all the children and adults who participate in the farm's education programs can learn. Such a change in perspective motivates our finding ways of acting that are rooted in a growing understanding and respect for the beings we are interacting with and affecting.

A major problem today is how distant most of us are, in terms of our awareness, from the effects of our actions. Think of all the products we use and consume. When I buy an inexpensive gallon of milk in a grocery story, I am most likely supporting factory farming and the whole economic system and worldview that drive it. I am basically saying with my purchase, without realizing it, "Produce more milk this way." Through my deeds I am connected with, and influencing, what happens in the world, but I may be oblivious to this fact.

This disconnect between myself as a consumer and the gifts I am consuming, which come from the earth, plants, animals, and toils of other human beings, is one consequence of the division of labor and an increasingly global economy. I am separated by countless steps from the larger reality and origins of the milk I drink or the clothes I wear. At the same time, my dependency on this complex web of relations becomes ever stronger. If I become conscious of this dependency, then a sense of gratitude arises for all that is given by other beings to make my life possible. At the same time, I can ask: how are those other beings and the earth being treated?

It is perhaps in response to the increased distancing effects of technology and of our economic system that the desire to connect consciously arises. What can I do to become aware of what I, concretely, am supporting through all my purchases? What policies and regulations can enhance the quality of life for animals and workers? Such questions have led to the pursuit of myriad activities. Think of the many, often intertwined, movements: animal rights, fair trade products, organic and regenerative agriculture practices, localization and regionalization of food production and distribution, and local currencies, to name a few. They all strive to consciously create connections that can allow all partners in the relations to thrive. There are many hurdles and no quick fixes. But the fact is that at the basis of these strivings lies a different way of knowing oneself embedded in the world.

Over two hundred years ago, Johann Gottfried Herder wrote, "The human being is the first to be set free in creation."[63] Animals are meaningfully woven into their contexts of life. (They do not pull back and start thinking about how they could make life better for themselves; they do not worry about whether they will make it through the coming winter or discuss strategies of how to do so.) When Herder said we are set free, he meant that we have become separate from the wise web of life inasmuch as we can ask questions, that we think about things as though from a distance, that we are uncertain about what the best ways to act may be, that we make many mistakes.

For the past centuries, Western culture—which is now present as a force around the globe—has thrived on trying to free itself from the bonds of nature. As a consequence, we have lost the wisdom that informs all life. We have banked on ingenuity. This only takes us so far, since smartness often leads to solutions that exploit or ignore the larger fabric of life on the planet. But we can turn for orientation toward the wisdom of life. That is also a gift of our being "set free."

CHAPTER 10

A Biology of Beings

The wolf does not become a lamb even if it eats nothing but lambs all its life. Whatever it is that makes it wolf, therefore, must obviously be something other than the "hyle," the sensory material, and that something, moreover, cannot possibly be a mere "thought-thing" even though it is accessible to thought alone, and not to the senses. It must be something active, something real, something eminently real.

I READ THIS PASSAGE from the nineteenth-century philosopher Vincenz Knauer for the first time years ago when I was in college.[1] It gave me pause to think then, and it still does so today. In one sense it is a straight-forward thought: Wolves eat lambs (and much else) and remain wolves; sloths eat mostly *Cecropia* leaves and remain sloths; frogs eat slugs and flies and remain frogs; lions eat zebras and remain lions; bison, zebras, and cows eat grass and remain bison, zebras, and cows. You see the point.

In a sense, all animals overcome their food to maintain themselves. And think of plants. Poppies, asters, and milkweeds, to name a few, all take in carbon dioxide, water, and some minerals and with the help of light create their own living substance and form. But how different they are from the water, air, and minerals they take in, and how different they are from one another!

Knauer's pondering points to the fact that organisms are activities. It is not the substance of the food that makes them what they are. It is the specific way of transforming and forming that makes the wolf a wolf, the frog a frog, the poppy a poppy.

We gain a most vivid sense of this creative activity in considering the development of an organism, as we did in the chapter on the frog. But the animal-as-activity does not cease to exist once it reaches adulthood. Certainly, there is more stability of form and substance in the adult. An adult frog carries out numerous activities—leaping with its long and powerful rear legs, catching a fly with its tongue, migrating to a vernal pool to mate, croaking at dusk. In all of these and many more activities we can point to a body and say, "That is a wood frog." In all the changing activity there is a certain stability of form (shape, size, color pattern, etc.).

While we can identify specific structures, the frog is always actively maintaining these and continually building up, breaking down, and transforming its bodily substances, all in relation to its needs and what it encounters in its surroundings. The frog never "is" in a static sense. The *mature* organism is also a being-at-work (to use again Joe Sachs' translation of Aristotle's term *energeia*). An organism is continually producing and maintaining itself. Its body is the momentary result of ongoing creative activity. As philosopher Susanne Langer puts it beautifully, "Every discovery makes the living organism look less like a predesigned object and more like an embodied drama of evolving acts, intricately prepared by the past, yet all improvising their moves to consummation."[2]

An animal as being-at-work is at work in the formation and main-tenance of all features of its body, and in all the activities of the body as a whole and its parts. The animal is a centered, integrated, and ever-engaged being. But while its body may be clearly distinct, an animal is not encapsulated in that body. A teacher of mine once remarked that the sun *is* where we find it at work. It is not just up there in the sky; it is present in the warming of my skin, in the greening and growing of the plants, in the melting of the snow, in the formation and movement of the clouds. So is an animal spread out into the world through all its activities and perceptions. This is its expansive nature. Each animal radiates out in a specific way—the lion in its way and the zebra in its. At the same time each animal is receptive to specific qualities in the world that allow it to develop and live. In this way animals and plants (and the other forces of nature) intermingle; they are part of each other and mold each other in their activities and relations. The planet is alive with interweaving and mutually evolving beings.

Gaining a sense of the integrated and expansive activity nature of any living creature is a first step or a first opening into a biology of beings. I don't use this expression without reticence, since I know that the term "being" is so easy to misunderstand and misconstrue. I am not appealing to some transcendent entity or to something "mystical" or vague hover-ing above the life of the animal. I am taking seriously the discovery that the animal is fundamentally integrative, creative activity and not a thing. The being of the animal I mean is being-at-work. The animal as activity is not a thing, but it is not nothing![3]

That is the key insight. Since every organism is activity, is being-at-work, its "nature" as an elephant or frog is not given as a thing. You can't place the "frogness" of the frog next to its liver, brain, heart, and stomach to examine it. You cannot point to it directly; it is not a spatial entity. And yet you can discover frogness in every detail of its anatomy, physiology, and behavior. But it is "in" all of these not as a thing to be found, but as effective agency, which we apprehend with our active mind that discovers the connectedness of all the particulars. A major problem is that our current modern habits of thought and our language tend toward spatiality. "In" implies physically within; "being" implies some discrete entity.

But these hurdles should not prevent us from forging new understanding. Knowledge evolves. Today some philosophers of biology and some biologists speak of organisms as processes, of agency and purposiveness, all of which are expressions that mainstream biology condemned 20 or 30 years ago.[4] So I am encouraged to go as far as I can and stick my neck out by using such phrases as being and being-at-work to characterize what I see to be essential in understanding animals.

In respect to the experiences of everyday life, what I am saying is not unknown or foreign. When we interact with a partner, children, grandchildren, colleagues, friends, or a person we strike up a conversation with at the gas station, we can be aware of the other as a specific presence. In our awareness this experience may range from a dull sense to a striking and luminous feeling of having met or connected with someone. Similarly, when I play with my dog, or follow two squirrels chasing each other around the trunk of a tree or sit in a meadow with ruminating cows, I can also sense unique presences. I have no trouble acknowledging the presence of other beings, and I know well that what I experience as a presence is—as a presence—not something I can grab onto or find thing-like in genes or the brain. A biology of beings is the endeavor to move from the tacit and often dim acknowledgement of the other beings, and to bring the unifying agency of those beings to greater and more differentiated appearance.

Scientific Portrayal

While there are no simple "steps," I can describe some important features of the scientific inquiry I am calling a biology of beings.

Riddle and Quest

Phenomena in the world appear to us as a riddle. How can I understand the giraffe's long neck? How can it be that the sloth is so slow? The huge elephant has such a flexible, multifunctional trunk: what a riddle! Without such ponderings and questions, there would be no impetus for scientific exploration. We want to connect in some way with what we feel disconnected from—it is therefore a riddle. And yet the experience of riddles and questions also reflects the fact that we have been touched by something in the world. There is a nascent connection that wants to grow. So we begin to explore. How this exploration plays out depends on the person and the conditions under which she or he is working.

If I want to express the fundamental impetus for my explorations, it is the question—as I mentioned in the chapter on the elephant—Who are you? This is a very open-ended question. It keeps me focused on the animal as a whole when I begin to delve into all its details. It is also respectful in that it acknowledges the other as a being in its own right. The intention of the quest is well stated by psychiatrist and phenomenologist Erwin Straus: "The knowledge [we] seek is not meant for controlling the world, but, rather, for unlocking it and letting a mute world become one that speaks to us in a thousand places."[5]

Engaging

As a researcher I carefully study the organism and work to gain an ever-better sense of its specific way of being. I notice and observe: The frog leaping into the pond when I come close; the frog floating with only its big bulging eyes and wide mouth breaking the water's surface; the varied colors of individual wood frogs; how differently tadpoles and adult frogs swim. So I attend to the frog. And I do not rely only on my own observations. I speak with others and read extensively in the scientific literature. Many people have dedicated their professional lives to studying myriad aspects of the animals I have portrayed, and I draw extensively from their findings and insights. In this sense I engage with a larger community and also am indebted to it.

Freeing

Because much biological research today is dedicated to discovering causes ("mechanisms") and to embedding findings in overarching theories (e.g., "evolution through natural selection") there is a good deal of thought work involved in trying to discern how findings are influenced or colored by frameworks. I work to free myself from contemporary biases and interpretations that constrict a more open-ended consideration of the phenomena. Conversations with colleagues is often a great help in this process. I do not want to place the facts into the context of a theoretical framework, but discover how they place themselves within the organism itself.

Picturing

By going into so many details, I may also begin to lose sense of the organism's way of being and its wholeness. I may lose the forest for the trees. So it takes constant effort to make conscious the connections and relations through which the organism reveals itself. To this end I try to picture what I am observing or the findings I am reading about as vividly as possible. I am not focusing in a narrow way on "why" an animal has this or that or does this or that. I am not trying to "explain" it. By vividly picturing the development of a tadpole into the adult frog I first realized that in a very essential sense it is not correct to say that the adult form develops "out of" a tadpole. By staying close to the observed phenomena and connecting the separate observations into a unity that reflects the unity that is the frog itself, I get a glimpse in thought of its way of being. The mental effort others put into theorizing I put into picturing.

Mutual Illumination

The particular way of being of an animal stands out all the more when we compare its characteristics with those of other animals. What it means to be a zebra becomes ever clearer when we compare it with a lion. Every species speaks its reality through its relations to others. So we need to let the different kinds of beings and their characteristics illuminate each other.

This is also the case with the various characteristics in one type of animal. I need to consider the giraffe's neck in relation to its short body, its long legs, its large perceptive eyes, the manifold ways in which it uses its neck, and of course much more. Only then do I have a chance of viewing "neck," which at the outset I may only grasp as a separate part, as an integral member of a unique being. No feature can be understood in isolation.

Insight and Wholeness

When I was in college and dissected a frog, I learned that it had a urostyle. At the time this bone made no big impression on me; unfortunately, it was simply one more part to memorize. In my recent study of the frog, the urostyle suddenly lit up. I no longer saw it as an isolated anatomical part but as a crucial member of this organism. I saw through it a quality of the frog. What is in other animals the extensive tail becomes in the tailless frog an internal bony structure that supports the strong leaping legs.

Many years ago, I was discussing the elephant with a twelfth-grade zoology class. I was beginning to see the manifold ways in which the elephant is a flexible giant—how it integrates what is delicate and sensitive with its power and robust nature. We came to its wrinkled, sparsely haired skin, and I described how the skin is so thick. Then I mentioned the surprising fact that through the skin an elephant can nonetheless detect a fly resting on it and twitch fine muscles, and the fly takes off. A student raised his hand and said something to the effect of, "Mr. Holdrege, that's just like all the other characteristics: the skin is thick and massive, but it is also so sensitive, just like the trunk." He had caught a glimpse of the whole through some parts. I had not really seen that connection before, and it was a high point of the class.

These examples show a form of perceiving meaning in the organism—how the "parts" are truly revelatory of the whole organism. This kind of intuiting is something you can no more make happen than you can make a frog appear in a pond. But you can prepare for such insights through all the work described above, so that you are moving in the territory in which connections can show themselves.

Portraying

In reality, any attempt to express the inner nature of a thing is fruitless. What we perceive are effects, and a complete record of these effects ought to encompass this inner nature. We labor in vain to describe a person's character, but when we draw together his actions, his deeds, a picture of his character will emerge.[6]

In one way it is surprising that Goethe—the giant of German literature and a very careful, methodical observer in his scientific work—would utter such words. But he had realized that the essential features of a person or any other being reveal themselves through the particulars and their relations. The character is not an abstract quality that you could define or nail down. So if you want to express the character—the holistic nature—of something, then you have the task to show how the different features reveal that character.

We can also think of a visual portrait in which the character of a person shines through the whole presentation and composition—through the way the parts are composed by the artist. He or she has glimpsed this character and seeks to give it expression.

A scientific portrayal of an organism is rather like that. In portraying I attempt to depict specific qualities, activities, and relations in such a way that the being-at-work of the animal can show itself to the reader. I can only suggest. As philosopher and author Owen Barfield points out, "Meaning itself can never be *conveyed* from one person to another; words are not bottles; every individual must intuit meaning for himself."[7] Since meaning expresses itself in relations, it can only speak between the lines in the active mind of the reader. Of course, much depends upon the felicity of expression and composition. If I succeed in describing the characteristics of an organism as vividly as possible and readers vividly picture what they read, then an understanding of the organism as a being can arise.

Encountering

When I have completed a portrayal I am not done, and my engagement with the animal is not something that I leave behind. After the intense process of working with a particular animal or plant, when I go out and see it in the wild again, my perception is enhanced. A green frog swimming in the pond is much more of a presence than it was before. The forceful and yet graceful kick of the legs, the shimmering green of its head, its bulging eyes—these details speak more strongly. My interest grows, along with a kind of elemental joy, in moments when I am able to participate in another being's way of being. I am more present for it to present itself. I then experience the truth of Emerson's statement: "It seems as if the day was not wholly profane, in which we have given heed to some natural object."[8]

*　*　*

When nature becomes a presence and I have been touched by another being, I also honor that presence, that being. This connection forms the basis for greater insight and, importantly, for an ethical relation to the natural world. A biology of beings fosters connection.

Acknowledgements

Most of the chapters of this book are based on articles and short monographs I have written over the years. Some time ago Beth Weisburn suggested that I put these studies into a book to make them more widely available. Her enthusiasm and support helped get the project off the ground.

I include new studies in the book and have reworked the content of all those earlier publications in the process of writing this book. I added recent observations and also perspectives that often came to me after doing a significant amount of new literature research. I am indebted to the many researchers past and present who have dedicated their lives to finding out more about animals and the nature of life. You can get a sense of the breadth of their efforts by perusing the book's references.

For anyone like me who is interested in the wisdom embodied in the skeletal structure of animals, the collections of museums provide an invaluable wealth of carefully prepared skeletal remains to study. I spent many hours over the years in such collections, mainly in the American Museum of Natural History in New York City, but also in the Field Museum in Chicago and the Academy of Natural Sciences of Drexel University in Philadelphia.

Trips to the western U. S., Africa, and South America allowed me to observe animals in their natural environments. I am still grateful to Wolfgang Schad, who, many years ago, encouraged me to take a trip to Africa to observe animals. He helped secure funding for that trip, which was in many ways a life changer. It was in all respects an eye-opening and thought-deepening experience to be with animals that previously I had known only through zoos, reading, and bones.

My thanks to friend and colleague Mark Riegner for coming up with the idea of co-leading a trip to the Amazon in 2015. The journey to this special part of the planet allowed me to be in the presence of sloths (and many other lovely creatures!), which was a great joy and led to new insights.

I'm especially grateful to my wife and colleague, Henrike Holdrege, for the many stimulating conversations about animals and the nature of reality we have had over the years. She also accompanied me out west to observe bison.

I'm fortunate to have the herd of dairy cattle at Hawthorne Valley Farm in immediate proximity, and the hours of sitting with the herd have been rewarding in many ways. Thanks to Hawthorne Valley farmers Spencer Fenniman and Steffen Schneider for sharing with me their practices and knowledge about cows and farming.

In talks, workshops, and courses I have had the opportunity to present the animals portrayed here. During preparation and in the courses often new insights also arise. This is the great value of "teaching"—to work together with people in explorations that lead to fresh perceptions and new understandings. So my thanks to the many participants in workshops and courses for productive co-working.

To my colleagues and friends who read and commented on parts or all of the manuscript, I say a hearty thanks. You helped a great deal: Linda Bolluyt, John Gouldthorpe, Henrike Holdrege, Bob Jensen, Seth Jordan, Elaine Khosrova, Kristy King, and Steve Talbott. Linda and Kristy also provided valuable assistance with notes and references.

I was fortunate to have Will Marsh as copy editor for the book; it's impressive what a careful eye and thoughtful mind discovers in a text. My thanks to Mary Giddens for doing the layout of the book.

The Nature Institute is a nonprofit organization that is funded through donations and grants. Both donations from individuals and grants from foundations have allowed me to carry out research and writing for this book. I am indebted to them all: Foundation for Rudolf Steiner Books, Mahle-Foundation, Rudolf Steiner-Fonds für wissenschaftliche Forschung, Salvia Foundation, Software AG Foundation, and the Waldorf Educational Foundation. The final phase of research and book writing was supported by the Evolving Science Association, a collaboration of The Nature Institute and the Myrin Institute.

Notes

Introduction

1. Goethe 1851. Translation by C. Holdrege.
2. Holdrege 2005a, 2013, and 2014.
3. Barfield 1965 and 1977; Steiner 1996 and 2011.
4. For Goethe's phenomenological approach to science see: Amrine et al. 1987; Bortoft 1996 and 2012; Goethe 1995; Richards 2002; Seamon and Zajonc 1998.
5. Here are some of the other philosophers and scientists with holistic approaches who have inspired my work: Kurt Goldstein (Goldstein 1995); Hans Jonas (Jonas 1966); Aldo Leopold (Leopold 1987); Paul Weiss (Weiss 1973).
6. I learned a great deal from and am deeply indebted to Jochen Bockemühl, Ernst-Michael Kranich, Wolfgang Schad, and Andreas Suchantke. Some of their work has been translated into English; see Bockemühl 1981; Kranich 1999; Suchantke 2009; Schad 2018.
7. The following articles, book chapters, and monographs written by me served—in whole or in part—as the basis for the following chapters in this book:
   Chapter 1: What Does it Mean to be a Sloth? In C. Holdrege & S. Talbott, *Beyond Biotechnology: The Barren Promise of Genetic Engineering* (pp. 132–153). Lexington: The University Press of Kentucky, 2008.
   Chapter 2: *The Flexible Giant: Seeing the Elephant Whole.* Ghent, NY: The Nature Institute, 2003.
   Chapter 3: How Does a Mole View the World? *In Context* #9, pp. 16–18, 2003.
   Chapter 5: Seeing the Animal Whole: The Example of Horse and Lion. In D. Seamon & A. Zajonc (Eds.), *Goethe's Way of Science* (pp. 213–232). Albany, NY: State University Press of New York Press, 1998.
   Chapter 6: Why Do Zebras Have Stripes? (Maybe it's the Wrong Question). *In Context* #37, pp. 17–23, 2017.
   Chapter 7: *The Giraffe's Long Neck: From Evolutionary Fable to Whole Organism.* Ghent, NY: The Nature Institute, 2005.
   Chapter 8: *Do Frogs Come from Tadpoles? Rethinking Origins in Development and Evolution.* Ghent, NY: The Nature Institute, 2017.
   Chapter 9: The Cow: Organism or Bioreactor? In C. Holdrege & S. Talbott, *Beyond Biotechnology: the Barren Promise of Genetic Engineering* (pp. 111–122). Lexington: The University Press of Kentucky, 2008.
8. Goethe 2002, p. 56. Translation by C. Holdrege.

Chapter 1 (Sloth)

1. Aiello 1985.
2. Sunquist 1986.
3. Castro-Vásquez et al. 2010; Sunquist and Montgomery 1973.

4. Mendel et al. 1985b.

5. I will be focusing mainly on a prominent species of three-toed sloth: the brown-throated sloth (*Bradypus variegatus*). See Hayssen 2010 for a description of this species.

6. Goffart 1971; Mendel 1985a.

7. Sunquist 1986, p. 9.

8. T. H. Bullock, quoted in Goffart 1971, p. 94.

9. Goffart 1971, p. 25.

10. Goffart 1971, p. 69.

11. Naples 1985.

12. Beebe 1926.

13. Mendel 1985a, p.159.

14. Varela-Lasheras et al. 2011.

15. Billet et al. 2012.

16. Goffart 1971, pp. 106ff.; Mendel et al. 1985; Piggins and Muntz 1985.

17. Beebe 1926.

18. Tirler 1966, p. 27.

19. Grzimek 1975.

20. Beebe 1926, p. 32.

21. Pauli et al. 2014.

22. Montgomery and Sunquist 1978.

23. Beebe 1926, p. 23.

24. Goffart 1971, p. 114.

25. Naples 1982; Hautier et al. 2016.

26. Naples 1982 p. 18.

27. Britton 1941.

28. Foley et al. 1995.

29. Montgomery and Sunquist 1978.

30. Urbani and Bosque 2007; Castro-Vásquez et al. 2010.

31. Beebe 1926.

32. The two-toed sloth defecates from the trees or on the ground, but since it has no tail, it leaves its feces lying on the leaf litter.

33. Montgomery and Sunquist 1975, p. 94.

34. Goffart 1971, p. 124.

35. Beebe 1926, p. 3.

36. Taube et al. 2001.

37. Montgomery, quoted in Sunquist 1986.

38. Goffart 1971, p. 59.

39. McNab 1978.

40. Gilmore et al. 2000.

41. Bourlière 1964.

42. Montgomery and Sunquist 1978.

43. Cliffe et al. 2015.

44. McNab 1978.

45. Waage and Best 1985; Gilmore et al. 2001.

46. Waage and Best 1985, p. 308.

47. Pauli et al. 2014.

48. Krakauer and Krakauer 1999.

49. Cited in Beebe 1926.

50. Beebe 1926, p. 13.

51. Goethe 1995, p. 121. Goethe wrote the essay that contains this statement in 1795; it was not published (in the original German) until 1820.

52. Goethe 1995, p. 121; translation modified by CH.

53. Bortoft 1996, p. 12.

Chapter 2 (Elephant)

1. Shoshani 1997.

2. Poole 1996, p. 138; see also Rasmussen and Munger 1996.

3. Moss and Colbeck 1992, p. 90.

4. Sukumar 1995, p. 52.

5. M. Krishnan, cited in Daniel 1998, p. 75

6. Owen-Smith and Chafota 2012.

7. Benedict and Lee 1938.

8. Rees 1982.

9. Moss 1988, p. 210.

10. Moss 1988, p. 217.

11. Quoted in Shoshani 1992, p. 132.

12. Rasmussen et al. 2002.

13. Siebert 2006.

14. Moss 1988, p. 152.

15. McComb et al. 2001.

16. Gobush et al. 2008.

17. Foley 2002.

18. Slotow et al. 2000.

19. Williams 1950.

20. Quoted in Daniel 1998, p. 73.

21. Sikes 1971, p. 82.

22. Tuskless elephants exist. One author comments that tuskless males "seem to make up in size of body and trunk for the lack of tusks and are generally enormously powerful elephants" (Evans, quoted in Daniel 1998, p. 61).

23. Schad 2018, pp. 937 ff.

24. Goethe 1995, p. 14.

25. Goethe 1995, p. 61.

26. Flindt 1986, p. 23.

27. Hutchinson et al. 2003.

28. Gambaryan 1974.

29. Sikes 1971, p. 32.

30. Kingdon 1989, p. 32.
31. Poole 1996, p. 156.
32. Sikes 1971, p. 78.
33. Kranich 1995, p. 161.
34. Sikes 1971, p. 54.
35. Chevalier-Skolnikoff and Liska 1993.
36. Kingdon 1989, pp. 46 and 58f.
37. Moss 1988, p. 270.
38. Sukumar 1995, p. 100.
39. Douglas-Hamilton et al. 2006.
40. McComb et al. 2005.
41. Poole 1996, pp.159f.
42. O'Connell-Rodwell 2007.
43. Payne 1989, p. 266; see also Payne et al. 1986 and Payne 1998.
44. Katherine Payne, quoted in Ben-Ari 1999.
45. O'Connell-Rodwell and Arnason 2000.
46. Reuter et al. 1998; see also O'Connell-Rodwell 2007.
47. For more related to sounds and communication, see: elephantvoices.org; Soltis 2010; Moss et al. 2011.
48. Poole et al. 2005.
49. For example, Schad 2018, pp. 937 ff.
50. Chadwick 1992, pp. 77f.
51. Herder 1982, pp. 64–5; translation C. Holdrege. This book was originally published in 1791; Herder lived from 1744 to 1803 and was a close friend of Goethe's.
52. Williams 1950, p. 78.
53. Poole 1996, p. 139.
54. Williams 1950, p. 78.
55. Poole 1996, p. 139.
56. In Shoshani 1992 p. 137.
57. Shoshani 1992, p. 137.
58. Williams 1950, p.78.
59. Shoshani 1992, p. 134.
60. Sukumar 1995, p. 50.
61. Haug 1970; see also Haug 1987.
62. Shoshani et al. 2006.
63. Flindt 1986, p. 108.
64. Haug 1970, p. 56.
65. Goethe 1995, p. 57.

Chapter 3 (Star-nosed Mole)
1. Nagel 1974.
2. Goethe 1995, p. 121.
3. Catania 2002, p. 56.

4. Grand et al. 1998.

5. Goethe 1995, p. 121.

6. Catania 2000, p. 66; see also Catania 2011.

7. Catania and Kaas 1995 and 1997.

Chapter 4 (Bison)

1. For descriptions of bison biology, see Allen 1876/1974; Lott 2002; McDonald 1981; McHugh 1979.

2. McHugh 1979, p. 22.

3. Lott 2002, p. 55.

4. McHugh 1979, p. 150.

5. McHugh 1979, p. 149.

6. Shaw 1995.

7. Hornaday 1889/2002, p. 525; Isenberg 2000.

8. Dodge 1877, p. 120.

9. Sanson 2006; Holechek 1984.

10. Sanson et al. 2007; McNaughton et al. 1985.

11. Jami et al. 2013.

12. For good reviews of the ecology of bison and the prairie, see Anderson 2006 and Knapp et al. 1999.

13. Fay 2003.

14. John Muir, July 27, 1869; in Muir 2003, p. 211.

15. Garretson 1938, p. 58.

16. Knapp et al. 1999.

17. Coppedge and Shaw 2000.

18. McMillan et al. 2011.

19. Gerlanc and Kaufman 2003.

20. Coppedge 2009; Jung et al. 2010.

21. Rosas et al. 2008.

22. Anderson 2006.

23. For the effects of fire and grazing on prairies today, see Fuhlendorf et al. 2009; Trager et al. 2004.

24. Stewart 2009.

25. Pyne 1988; Whitney 1996.

26. Quoted in Stewart 2009, p. 136.

27. Quoted in Stewart 2009, p. 115.

28. McHugh 1979, chapter 8.

29. McHugh 1979, p. 99.

30. McHugh 1979, p. 105.

31. McHugh 1979, p. 99.

32. There is a wealth of literature on the intertwined spiritual and physical culture of Native Americans; see, for example, Brown 1989 and 2007; DeMallie 1985; Irwin 1996; Linderman 2002; Posthumus 2016; Walker 1980.

33. Grinnell 1962, pp. 125–6.
34. Grinnell 1962, p. 125.
35. Kehoe 1965; Zedeño 2008.
36. Brown 1989, p. 9.
37. Brown 1989, pp. 5–6.
38. Brown 1989.
39. Brown 1989, p. 72.
40. Grinnell 1962, p. 101.
41. Skinner 1915, p. 710.
42. Isenberg 2000; Flores 1991; Flores 2016, Chapter 6.
43. DeMallie 1985, p. 290.
44. Linderman 2002, p. 169.
45. Gates et al. 2010.
46. Bailey 2013.
47. Gates et al. 2010; Freese et al. 2007; Sanderson et al. 2008.
48. http://www.ncai.org/conferences-events/ncai-events/Land_and_Natural_
    Resources_Committee_-_Inter_Tribal_Buffalo_Council_Presentation.pdf
49. Flores 2016, p. 134–5.
50. Jonas 1968, p. 233.

Chapter 5 (Zebra and Lion)

1. Groves and Bell 2004.
2. Estes 1991, p. 236.
3. McNaughton 1985.
4. Estes 1991, Sanson 2006.
5. Brooks and Harris 2008.
6. Bell 1971.
7. Naidoo et al. 2016
8. Moss 1982, pp. 99 f. and 105.
9. Estes, 1991, p. 237.
10. Klingel 1969 and 1972.
11. Hayward and Slotow 2009; Schaller 1972.
12. Moss 1982, p. 255.
13. Lindbergh 1966.
14. Mosser and Packer 2009.
15. Schaller 1972, p. 157.
16. Hayward and Kerley 2005; Sinclair et al. 2003.
17. On lion hunting: Fischhoff et al. 2007; Haas, 2005 (has many references); Sinclair
    et al. 2003; Funston et al. 2001; Funston et al. 1998; Elliott et al. 1977; Schaller 1972.
18. Schaller 1972, p. 268.
19. Klingel 1966; Wackernagel 1965.
20. Estes 1991, p. 376.
21. Smuts 1975.

22. Budiansky 1997, Gray 1968.
23. Nickel et al. 1986.
24. Budiansky 1997; Hildebrand 1974.
25. Estes 1991, p. 374.
26. Pfefferle et al. 2007.

Chapter 6 (Zebra Stripes)

1. Portmann 1967.
2. Portmann 1967, Chapter XI.
3. Schad 2018; Riegner 1998.
4. Cabrera 1936; Suchantke 2001.
5. Suchantke 2001, pp. 5–9.
6. Ruxton 2002.
7. Larison et al. 2015a.
8. Ruxton 2002; Larison et al. 2015a; Caro et al. 2014; Caro 2016.
9. See, for example, Ruxton 2002; Larison et al. 2015a and 2015b; Caro et al. 2014; Caro 2016; Horváth et al. 2018.
10. Caro 2016, p. 23.
11. Melin et al. 2016.
12. Elliott et al. 1977; Funston et al. 2001.
13. Morris, cited in Ruxton 2002, p. 238.
14. Caro 2016, p. 193.
15. Caro et al. 2014.
16. Larison et al. 2015b.

Chapter 7 (Giraffe)

1. The sausage tree (*Kigelia africana*), a member of the Jacaranda family (Bignoniaceae, to which trumpet vine and catalpa belong), grows in south-central and southeastern African.
2. Jean-Baptiste Lamarck (1744–1829); see Lamarck 1984, p. 122.
3. Lamarck 1984, p. 45.
4. Charles Darwin (1809–1882); see Darwin 1872, pp. 177 ff.
5. See, for example, Jablonka and Lamb 1995; Jablonka and Lamb 2005; Jablonka and Raz 2009.
6. Ciofolo and Le Pendu 2002.
7. Cameron and du Tuit 2007.
8. Young and Isbell 1991; Ginnett and Demment 1997; Woolnough and du Toit 2001; Shorrocks 2009.
9. Ginnett and Demment 1997.
10. Leuthold and Leuthold 1972 and 1978; Pellew 1984.
11. Mitchell et al. 2010.
12. Gould 1996; Mitchell and Skinner 2003, p. 69.
13. El Aich et al. 2007.

14. Simmons and Scheepers 1996.
15. Pincher 1949.
16. Study by U. de. V. Pienaar; cited in Simmons and Scheepers 1996.
17. Brownlee 1963.
18. Assuming for the sake of explanation a spherical body, the two-dimensional surface grows as a function of the square of the radius, while the volume—being three-dimensional—grows as a function of the cube of the radius. Therefore the proportion of the volume to surface area grows as a function of the radius. For example, a sphere with a radius of 2.5 cm (about one inch) has a volume-to-surface ratio of 0.8:1. A much larger sphere with a radius of 50 cm (about 20 inches) has a volume-to-surface ratio of 16.7:1.
19. See Mitchell et al. 2017; Krumbiegel (1971) estimated that the ratio of volume to surface in the giraffe is 11:1, compared, say, to a smaller, long-necked antelope, the gerenuk, which has a ratio of 4.7:1 (similar to the human).
20. Brownlee 1963; Williams 2016.
21. Cameron and du Toit 2005.
22. Wilkinson and Ruxton 2012.
23. Simmons and Scheepers 1996; Simmons and Altwegg 2010.
24. Dobzhansky 1973.
25. Quoted in Spinage 1968 , p. 41.
26. Quoted in Spinage 1968, p. 54.
27. Spinage 1968, p. 85. Today there is much discussion about whether there is one giraffe species with a variety of subspecies or whether there are multiple species. I will not address here this "taxonomic quandary"; for a brief review see Bercovitch and Deacon 2015.
28. Darwin 1872, p. 177.
29. van Schalkwyk et al. 2004.
30. Stephen Jay Gould (1996) and Mitchell and Skinner (2003) claim that the giraffe's forelegs only appear to be longer than the hind legs. But when one measures the lengths of the individual bones of the forelegs and the hind legs, the forelimbs are clearly longer (see Table 1 below, and also Colbert 1935). It is not a matter of "mere appearance." The tendency toward lengthening is stronger in the front part of the body.
31. cf. Kranich 1995 , pp. 138–46.
32. See Table 2 below.
33. Lankester 1908.
34. Solounias 1999.
35. Owen 1839, p. 234.
36. Murie 1872.
37. See Slijper 1946; and Table 3.
38. Goethe 1995, pp. 120–21.
39. See Table 3; body length is measured as the sum of the length of the thoracic and lumbar regions of the spine.

Source of Data	Harris (1976)	Holdrege (2005b)
Sample Size	14 specimens from Kenya National Musuem, Nairobi	7 specimens from the American Museum of Natural History
Sex	male	male, female, unknown
Length of forelimb bones (mean and range) (humerus, radius, metacarpus)	210 cm (range not reported)	192 cm (range: 177 to 210)
Length of hind limb bones (mean and range) (femur, tibia, metatarsus)	196 cm (range not reported)	180 cm (range: 167 to 193)
Length difference (mean and range)	14 cm (range not reported)	12 cm (range: 6 to 17)
Ratio of forelimb length to hind limb length	107.1 : 100	106.7 : 100

Table 1. These data show that the giraffe's forelegs are longer than its hind legs. The length of the three longest bones in the forelimb and hind limb were measured in a total of 21 specimens; in all cases the forelimb was longer than the hind limb.

(*The data from Harris 1976 were modified to give the length of the radius and not the combined length of the radius and ulna, which he reports. The proximal end of the ulna extends on average 11 cm beyond the proximal end of the radius [my data], which articulates with the humerus. This amount was subtracted from his reported radius/ulna lengths in order to give a truer picture of overall limb length.)

Species	Length of neck relative to length of forelimb (forelimb = 100)	Length of neck relative to length of hind limb (hind limb = 100)
Elk	66.4	55.2
Okapi	62	61.4
Giraffe	84.7	91.3

Table 2. Neck length (seven cervical vertebrae) is given as a percentage limb length (three longest bones of each limb, excluding phalanges). The giraffe's neck is proportionately much longer in relation to the limbs than that of the okapi and elk, even though the giraffe has longer legs than any other mammal.

(Note: Sample size was seven specimens of each species; each specimen was measured and the mean length in each group of seven was used to establish the ratios; from Holdrege 2005b.)

Species	Length of neck relative to length of forelimb (forelimb = 100)	Length of neck relative to length of hind limb (hind limb = 100)
Bison	93.5	113.8.
Elk	121.5	151.1
Giraffe	225.5	210.0

Table 3. Length of the limbs as a percentage of body length (thoracic/lumbar spine); i.e., length of the body equals 100. The bison's limbs are about the same length as its body, while the giraffe's are more than twice the length of its body. (Data from Slijper 1946)

40. Owen 1841, pp. 219–220.
41. Dawkins 1996, p. 103.
42. Mitchell and Skinner 1993.
43. See Table 4.
44. Kimani and Mungai 1983; Kimani and Opole 1991.
45. van Critters et al. 1966; Warren 1974.
46. Hargens et al. 1987.
47. Mitchell and Skinner 1993.
48. Mitchell and Skinner 2009; Østergaard et al. 2013.
49. König 1983.
50. Mitchell et al. 2013.
51. Backhaus 1959.
52. Stevens 1993, p. 10.
53. Krumbiegel 1971, p. 52.
54. Baotic et al. 2015; Bercovitch and Deacon 2015.
55. von Muggenthaler et al. 1999.
56. Stevens 1993, p. 6.
57. Dagg and Foster 1982, p. 102.
58. Langman 1982, p. 96.
59. Estes 1991, p. 205.
60. Coe 1967.
61. Estes 1991, p. 203.
62. VanderWaal et al. 2014.
63. Reported in Milius 2003.
64. VanderWaal 2014; Bercovitch and Berry 2012; Bercovitch and Deacon 2015.
65. Dagg and Foster 1982, p. 65.
66. See for example, Bercovitch and Berry 2012; Estes 1991 (pp. 201–7); Langman 1982; Pratt and Anderson 1979.
67. Fennessy et al. 2003.
68. Fennessy et al. 2003.
69. Hofmann and Matern 1988.
70. Pellew 1984.
71. Owen 1839, p. 219.
72. Dagg and Foster 1982, p. 82.
73. Hofmann 1973, p. 119.
74. Hofmann 1973, p. 119.
75. Milewski et al. 1991.
76. Hofmann and Matern 1988.
77. Milewski et al. 1991.
78. Young and Okello 1998.
79. Madden and Young 1992.
80. Du Toit 1990a.
81. McNaughton 1983.

Height (in meters)	Hydrostatic pressure in a water column (mm Hg)	Mean arterial pressure in Giraffe (mm Hg)	Mean venous pressure in Giraffe (mm Hg)
4: level of head	0		
3.5: upper neck	40	100 (±21)	16 (jugular vein)
3			7 (jugular vein)
2: heart level	150	185 (±42)	0 (right atrium)
1: ankle joint			
0: feet	300	260 (range 70 to 380)	150 (range –250 to +240)

Table 4. Blood Pressure in the Giraffe.
This table shows that the giraffe's blood pressure varies significantly depending on the height of the body part in which the measurements are taken (left column) and whether the blood was flowing through an artery (third column) or a vein (fourth column). The pressures in a four-meter-high standing column of water are given as a comparison (second column).

Scientists generally express blood pressure in millimeters of mercury (mm Hg). Mercury, a metal, is a very dense fluid at room temperature, so the pressure of 1 mm Hg is equal to that of a 13.6 mm high column of water. If you want to know the pressure at the bottom of a column of water expressed in mm Hg, you measure the height in millimeters and divide the number by 13.6. Most of us are used to stating blood pressure with two numbers, such as 120/80, which is a typical blood pressure in a resting human, with the first number expressing the higher pressure when the left ventricle contracts (systolic pressure) and the lower number expressing the lower pressure when the left

ventricle relaxes (diastolic pressure). Scientists often use, for simplicity's sake, only one number, the mean arterial pressure, which is simply halfway between systolic and diastolic pressures.

Even taking into account that the blood pressure in a resting animal can vary greatly, measurements of blood pressure in giraffes suggest that the mean pressure is significantly higher than in most other mammals. The mean pressure at the level of the heart in a standing giraffe is 185 mm Hg (±42; Mitchell and Skinner 1993). In contrast, the mean arterial pressure (resting) in mammals as different as cattle, dogs, and mice is between 100 and 120 mm Hg. In humans the average mean pressure is 100 mm Hg.

Note the large blood pressure oscillations in the giraffe's feet, which arise when it walks and runs. Arterial pressure in the feet can go from 70 to 380 mm Hg while the venous pressure in the feet varies between minus 250 (!) and plus 240 mm Hg. These extreme changes pose a major riddle. (Sources for data: Hargens et al. 1987, Mitchell and Skinner 1993.)

82. Bond and Loffell 2001.
83. Du Toit 1990b.
84. Kranich 1995 and 1999.
85. Eisely 1961; Teichmann 1989.
86. Quoted in Teichmann 1989, pp.13–14; transl. CH.
87. Here are a few examples: Berg 1922/1969; Bergson 1911/1998; Bowler 2003; Goodwin 2001; Goodwin and Webster; 1996; Gutmann 1995; Ho and Saunders 1984; Kranich 1999; Levit et al. 2008; Riedel 1978; Turner 2017; Verhulst 2003; see also see www.thethirdwayofevolution.com.
88. Dembski and Wells 2008; Meyer 2004.
89. Spinage 1968, p. 153.
90. Lindsey et al. 1999, Hart and Hart 1989.
91. I draw here primarily from Badlangana et al. 2009; Bohlin 1926; Churcher 1978; Colbert 1938; Harris 1976; and Mitchell and Skinner 2003; Danowitz et al. 2015a.
92. Bohlin 1926.
93. Churcher 1978; Danowitz et al. 2015b.
94. Cited in Churcher 1978.
95. Churcher 1978, p. 514.
96. Schindewolf 1969; see also Schindewolf 1950/1993.
97. Eldredge and Gould 1972; Gould and Eldredge 1993; see also Gould 2002.
98. Quotation from Carroll 1997, p. 167.
99. Lampi et al. 1992.
100. Schad 2018; the first German edition of this groundbreaking work was published in 1971. See also Riegner 1998.
101. Suchantke (2009) gives examples within different groups of animals; Riegner describes the pattern in mammals (1998) and birds (2008); Lockley (1999, 2007)) gives examples in dinosaurs.
102. See, for example, Whitham et al. 2003.
103. Mitchell and Skinner 2003.
104. Goethe 1995, pp. 55f.
105. Mitchell and Skinner 2003, p. 64.
106. Eisely 1961, pp. 341–342.
107. This is changing; see, for example, http://www.thethirdwayofevolution.com and the work of my colleague Stephen Talbott (http://natureinstitute.org/txt/st/org/index.htm).

Chapter 8 (Frog)

1. Duellman & Treub 1994, p. 365.
2. I am limiting my descriptions to those tadpoles that develop into frogs (not the tadpoles that develop into salamanders). I will focus on pond-dwelling tadpoles and their metamorphosis into land-dwelling frogs, as exemplified by many species that live in temperate climates. There is an astounding variety of ways

in which different species of frogs develop—some have no tadpole phase, some have tadpoles that are carnivorous rather than herbivorous, some remain aquatic for their entire life cycle, and so on. Because of this, for every characteristic I describe there are probably exceptions. They are fascinating and warrant consideration when you really want to understand the peculiarities of given species or genera and the variations within the amphibians. But my aim here is to provide a general picture of metamorphosis.

3. I have drawn on the following literature for my description of frog biology and metamorphosis: Brown and Cai 2007; Dodd and Dodd 1976; Duellman and Trueb 1994; McDiarmid and Altig 1999; Shi 2000.

4. Thinkers who have a keen awareness of thought and strive to keep their thinking close to the phenomena have remarked on the impossibility of deriving what comes later from what came earlier. See, for example, Goethe 1995, p. 8; Steiner 2011, Chapter 12; Goldstein 1995, pp. 374–5).

5. Huxley 1860.

6. See the historical overview in Keller 2008.

7. Sachs 2005.

8. For a detailed consideration, see Talbott 2015. In addition, if you are interested in gaining an impression of the overwhelming amount of research showing "How the Organism Decides What to Make of Its Genes," see Talbott's extensive collection of notes: http://www.natureinstitute.org/txt/st/org/support/genereg.htm

9. Reported in de Almeida and Carmo-Fonseca 2012.

10. From Meyer et al. 2013.

11. In his classic work, *The Organism,* Kurt Goldstein showed how in the healthy organism all activities of "parts" are not autonomous, but performances of the whole organism. His book provides a truly organic view of the living organism. See Goldstein 1995.

12. Gilbert 2016.

13. It is possible to expand the concept of causation beyond the narrow notion of mechanistic causality that reigns in conventional biological explanations. Aristotle spoke of four kinds of causes, and what I have been pointing to relates primarily to what he calls "formal" causation (see Aristotle 1995). Kant speaks of the organism as "both cause and effect of itself" (Critique of Judgment 1951, § 64; p. 217). See also Hans Jonas 1966, pp. 79 ff.

14. Pfennig 1992; Ledón-Rettig and Pfennig 2011; Storz et al. 2011.

15. Pfennig 1992.

16. Storz et al. 2011.

17. Bortoft 2013.

18. Levit et al. 2008 and Bowler 2003.

19. The website http://www.thethirdwayofevolution.com/ links to many articles.

20. See, for example, Roček and van Dijk 2006.

21. Shubin and Jenkins 1995.

22. Roček 2000, p. 1295.
23. Carroll 2000, p. 1270.
24. Roček and Rage 2000.
25. For articles about *Czatkobatrachus* see: Evans and Borsuk-Bialynicka 2009; Roček and Rage 2000.
26. Anderson et al. (2008) describe the fossil *Gerobatrachus hottoni*.
27. *Acanthostega* is described in Clack 2006, and in Clack 2012, pp. 161–174.
28. McFadden 1999.
29. For example, Lordkipanidze et al. 2013 and Lovejoy et al. 2009.
30. Whitehead 1967, Chapter IV.
31. Duellman and Trueb 1994, p. 425.
32. Cartmill and Smith 2009, p. 62.
33. Romer 1966, p. 25.
34. Carroll 1997, p. 152.
35. Romer 1966, p. 88.
36. Eleven articles in the October 2, 2009, issue of *Science* (vol. 326) are concerned with *Ardipithecus*.
37. Lovejoy et al. 2009.
38. See, for example, Schulz 1930; Schulz 1936; Straus 1949; Sayers et al. 2012.
39. Lovejoy et al. 2009.
40. White et al. 2015.
41. The realization that you can't determine a specific ancestor has led some evolutionary scientists to practice what is called pattern cladistics (Brady 1985 and 1994; Ebach and Williams 2019; Williams and Ebach 2009). They try to establish relationships between forms: those forms that share the greatest number of specialized ("derived") characteristics are considered to be most closely related to each other. These researchers try to construct dichotomously branching diagrams that express greater and lesser relationship in the context of time. They do not speculate—when they stay true to their principles, which is not always the case—about which forms evolved from which. This is a positive development inasmuch as it restrains speculation and focuses attention on relations and patterns that actually can be observed and discerned by comparing fossils. Problematic in cladistics more generally is the tendency to dissolve organisms into collections of individual specialized traits, and solely on this basis to establish relatedness that is supposed to underlie the evolution of real-life cohesive organisms.
42. I'd like to point to work carried out by a number of different scientists who have discussed overriding and often surprising patterns in morphology and evolution. These are stimulating contributions to gain a more dynamic and broader view of evolutionary processes that are not limited by the standard explanatory frameworks. See Kranich 1999; Lockley 2007; Riegner 2008; Rosslenbroich 2014; Schad 2018; Suchantke 2009. For additional references, see Holdrege 2009.

Chapter 9 (Dairy Cow and Responsibility)

1. Swenson 1990, p. 292. This volume contains much valuable information about cow digestion and physiology.
2. Kranich 1995, pp. 19–29.
3. Isaac 1962.
4. Zeder 2011.
5. Ajomone-Marsan and Lenstra 2010; Bollongino et al. 2012.
6. Hahn 1896; Harlan 1998; Isaac 1962; Cauvin 2000; Russell 2011.
7. https://en.wikipedia.org/wiki/List_of_cattle_breeds
8. Gepts and Papa 200; Price 1984 and 1999; Zeder 2012.
9. Amaral-Phillips et al. 2006.
10. See: https://www.nass.usda.gov/Charts_and_Maps/Milk_Production_and_Milk_Cows/cowrates.php and also http://www.holsteinusa.com/pdf/fact_sheet_cattle.pdf
11. In 1944, there were 25.6 million dairy cows in the U.S and each produced an average of 532 gallons of milk per year. In 2012, by contrast, there were only 9.3 million cows and each produced an average of 2,526 gallons of milk per year. Total milk production rose from 117 billion pounds in 1944 to from to 200.6 billion pounds in 2012 (https://nass.usda.gov/Publications/Trends_in_U.S._Agriculture/Livestock_and_Dairy/index.php; and https://www.agcensus.usda.gov/Publications/2012/).
12. U.S. Department of Agriculture 2009.
13. U.S. Department of Agriculture 2016.
14. https://www.ucdavis.edu/news/how-genetic-mutation-1-bull-caused-loss-half-million-calves-worldwide/ and http://www.newsweek.com/2016/11/25/dairy-cows-cattle-genetic-mutation-521152.html
15. U.S. Department of Agriculture 2016.
16. Krause and Oetzel 2006; National Research Council 2001.
17. U.S. Department of Agriculture 2016.
18. McKenna 2017.
19. U.S. Department of Agriculture 2014b (see table G.1b., p. 149); U.S. Department of Agriculture 2017.
20. Fulwider et al. 2008.
21. "Overall, U.S. milk consumption on average is rising between 1 and 2 percent annually but production is going up around 3 percent." See: https://www.cnbc.com/2017/09/22/dairy-glut-in-us-leads-to-problem-of-spilled-milk.html
22. http://ageconsearch.umn.edu/record/33612/files/ai010761.pdf
23. https://www.usda.gov/media/press-releases/2016/08/23/usda-purchase-surplus-cheese-food-banks-and-families-need-continue
24. https://www.marketwatch.com/story/got-milk-too-much-of-it-say-us-dairy-farmers-2017-05-21
25. Gardner 2002, p. 220.
26. In 1939 there were 21.9 million dairy cows in the U.S. on a total of 4,663,431 farms. 99% of the farms had 30 or fewer cows, while only 0.2% had 50 or more cows.

The USDA census stopped listing size at "50 or more." http://usda.mannlib. cornell.edu/usda/AgCensusImages/1964/02/02/743/Table-15.pdf

27. U.S. Department of Agriculture 2014a.
28. U.S. Department of Agriculture 2016.
29. Brickel and Wathes 2011; Espejo et al. 2006; Hadley et al. 2006; Hansen 2000; Hare et al. 2006; Knaus 2009; Lucy 2001; Oltenacu and Broom 2010; Walsh et al. 2011; Wathes 2012.
30. Russell and Rychlik 2001.
31. Plazier et al. 2008.
32. Walsh 2011.
33. Sordillo et al. 2009.
34. Van Boeckel et al. 2015; U.S. Department of Agriculture 2016. Globally, 73 percent of all antibiotics sold are given to animals that are raised for food (Van Boeckel et al. 2019).
35. Heuer et al. 2011; Marshall and Levy 2011.
36. http://www.bclaws.ca/civix/36/id/consol14/consol14/96044_pit
37. See Eicher et al. 2006; Humane Society of the United States 2012; Stafford and Mellor 2011.
38. Gottardo et al. 2011.
39. Knierim et al. 2015.
40. Fulwider et al. 2008; Sutherland and Tucker 2011.
41. Sutherland and Tucker 2011.The practice of tail docking in dairy cows may be on the way out in the U.S. Although still allowed at the federal level, it was forbidden in California in 2009. The American Veterinary Medical Association opposes routine docking of tails in dairy cows and the National Milk Producers Federation—whose members produce the bulk of milk in the U.S.—required members to begin phasing out routine tail docking at the beginning of 2017 (https://www.avma.org/KB/Resources/LiteratureReviews/Pages/Welfare-Implications-of-Tail-Docking-of-Cattle.aspx?PF=1; https://www.avma.org/KB/Policies/Pages/Tail-Docking-of-Cattle.aspx; http://www.nmpf.org/files/Tail%20 Docking%20Release%20TB%20102615.pdf).
42. Riegner 1998; Schad 2018, chapters 7 and 8; Spengler Neff et al. 2016.
43. Solounias et al. 1995.
44. Sisson and Grossman 1953, p. 144; Nickel et al. 1986, p. 157.
45. Probst et al. 2017.
46. Knierim et al. 2015.
47. Irrgang 2012.
48. Shriver 2009.
49. Schultz-Bergin 2017.
50. Fulwider et al. 2008.
51. https://www.cornucopia.org/scorecard/dairy/
52. U.S. Department of Agriculture 2016.
53. Kilgannon 2018.

54. McIntosh et al. 2016.

55. In October 2018, Hawthorne Valley Farm sold its premium raw milk to its dairy for .$42 per pound, while the wholesale price for conventional milk was only .$16 per pound.

56. For other perspectives on animal sentience and transforming our relation to animals see, for example, Abram 2011; Bekoff 2002; Sloan 2015.

57. Leopold 1987, pp. 129–30. The book was originally published in 1949.

58. For a more comprehensive consideration of Leopold's encounter with the wolf and the subsequent transformation of his worldview, see Holdrege 2016.

59. Leopold 1987, p. 130.

60. Leopold 1987, p. 130.

61. Leopold 1987, p. 131.

62. Leopold 1987, p. 204.

63. Herder 1982. This book was originally published in 1791; Herder lived from 1744 to 1803 and was a close friend of Goethe's.

Chapter 10 (Biology of Beings)

1. This is my translation of Knauer's German text and is in a chapter concerned with Aristotle's philosophy (Knauer 1892 p. 137).

2. Langer 1967, vol. 1, p. 378.

3. I borrow this expression from Henri Bortoft; see Bortoft 1996, pp. 13ff.

4. See the writings of my colleague, Stephen Talbott, on a "Biology Worthy of Life" (http://natureinstitute.org/txt/st/org/index.htm) and the following books: Nicholson and Dupré 2018; Walsh 2015.

5. Straus 1963, p. 395

6. Goethe, 1995, p. 158.

7. Barfield 1973, p. 133; his emphasis.

8. From Emerson's 1844 essay "Nature"; in Emerson 1983, p. 542.

References

Abram, David (2011). *Becoming Animal*. New York: Vintage Books.

Aiello, Annette (1985). Sloth Hair: Unanswered Questions. In: G. Gene Montgomery, Ed., *The Evolution and Ecology of Armadillos, Sloths, and Vermilinguas*. Washington, D. C.: Smithsonian Institution Press, pp. 213–8.

Ajmone-Marsan, Paolo et al. (2010). On the Origin of Cattle: How Aurochs Became Cattle and Colonized the World, *Evolutionary Anthropology: Issues, News, and Reviews* vol. 19, pp. 148–57.

Allen, J. A. (1876/1974). *The American Bisons Living and Extinct*. New York: Arno Press.

Amoroso, E. C. et al. (1947). Venous Valves in the Giraffe, Okapi, Camel and Ostrich, *Journal of Zoology* vol. 117, pp. 435–40.

Amrine, Frederick et al., Eds. (1987). *Goethe and the Sciences: A Reappraisal*. Dordrecht, Holland: D. Reidel Publishings Company.

Amaral-Phillips, Donna M. et al. (2006). Feeding and Managing Baby Calves from Birth to 3 Months of Age. University of Kentucky Cooperative Extension Service.

Anderson, J. et al. (2008). A Stem Batrachian from the Early Permian of Texas and the Origin of Frogs and Salamanders, *Nature* vol. 453, pp. 515–18. doi:10.1038/nature06865.

Anderson, Roger C. (2006). Evolution and Origin of the Central Grassland of North America: Climate, Fire, and Mammalian Grazers, *The Journal of the Torrey Botanical Society* vol. 133, pp. 626–47.

Aristotle (1995). *Aristotle's Physics: A Guided Study*. Joe Sachs (translator). New Brunswick, NJ: Rutgers University Press.

Backhaus, D. (1959). Experimentelle Prüfung des Farbsehvermögens einer Masai–Giraffe, *Ethology* vol. 16, pp. 468–77.

Badlangana, Ludo et al. (2009). The Giraffe (*Giraffa camelopardalis*) Cervical Vertebral Column: a Heuristic Example in Understanding Evolutionary Processes? *Zoological Journal of the Linnean Society* vol. 155, pp. 736–57.

Bailey, James A. (2013). *American Plains Bison: Rewilding an Icon*. Helena, MT: Sweetgrass Books.

Baotic, Anton et al. (2015). Nocturnal "humming" Vocalizations: Adding a Piece to the Puzzle of Giraffe Vocal Communication, *BMC Research Notes* vol. 8, pp. 425 ff.

Barfield, Owen (1965). *Saving the Appearances*. New York: Harcourt, Brace & World.

——— (1973). *Poetic Diction* (3rd edition). Middletown, CT: Wesleyan University Press.

——— (1977). *The Rediscovery of Meaning*. San Rafael, CA: The Barfield Press.

Beebe, William (1926). The Three-toed Sloth, *Zoologica* vol. VII, pp. 1–67.

Bekoff, Marc (2002). *Minding Animals*. New York: Oxford University Press.

Bell, Richard (1971). A Grazing Ecosystem in the Serengeti, *Scientific American* vol. 225(1), pp. 86–93. doi: 10.1038/scientificamerican0771-86.

Benedict, Francis G. and R. C. Lee (1938). Further Observations on the Physiology of the Elephant, *Journal of Mammology* vol. 19, pp. 175–94.

Ben-Ari, Elia T. (1999). A Throbbing in the Air, *BioScience* vol. 49, pp. 353–8.

Bercovitch, Fred and Philip Berry (2012). Herd Composition, Kinship and Fission-fusion Social Dynamics among Wild Giraffe, *African Journal of Ecology* vol. 51, pp. 206–16.

Bercovitch, Fred and Francois Deacon (2015). Gazing at a Giraffe Gyroscope: Where Are We Going? *African Journal of Ecology* vol. 53, pp. 135–46.

Berg, Leo S. (1926/1969). *Nomogenesis or Evolution Determined by Law*. Cambridge, MA: M.I.T. Press.

Bergson, Henri (1911/1998). *Creative Evolution*. Mineola, NY: Dover Publications. (This is a reprint of the 1911 translation; the French original was published in 1907.)

Billet, Guillaume et al. (2012). High Morphological Variation of Vestibular System Accompanies Slow and Infrequent Locomotion in Three-toed Sloth, *Proceedings of the Royal Society of London B: Biological Sciences* vol. 279, pp. 3932–9.

Bockemühl, Jochen (1981). *In Partnership with Nature*. Wyoming, RI: Bio-dynamic Literature.

Bohlin, Birger (1926). *Die Familie Giraffidae mit besonderer Berücksichtigung der fossilen Formen aus China*. Peking: Geological Survey of China (*Palaeontologia Sinica Series C* vol. IV [Fascicle 1]).

Bond, William J. and Debbie Loffell (2001). Introduction of Giraffe Changes Acacia Distribution in a South African Savanna, *African Journal of Ecology* vol. 39, pp. 286–94.

Bollongino, Ruth et al. (2012). Modern Taurine Cattle Descended from Small Number of Near-Eastern Founders, *Molecular Biology and Evolution* vol. 29, pp. 2101–4.

Bortoft, Henri (1996). *The Wholeness of Nature*. Hudson, NY: Lindisfarne Press.

Bortoft, Henri (2012). *Taking Appearance Seriously*. Edinburgh: Floris Books.

———— (2013). The Form of Wholeness, *In Context* # 29, pp. 8–11. Available online at: http://www.natureinstitute.org/pub/ic/ic29/bortoft.pdf.

Bourlière, François (1964). *The Natural History of Mammals*. New York: Alfred A. Knopf.

Bowler, P. J. (2003). *Evolution: The History of an Idea*. Berkeley: University of California Press.

Brady, Ronald H. (1985). On the Independence of Systematics, *Cladistics* vol. 1, pp. 113–26. Available online at: http://natureinstitute.org/txt/rb

——— (1994). Pattern Description, Process Explanation, and the History of the Morphological Sciences. Originally published in: L. Grande and O. Rieppel, Eds., *Interpreting the Hierarchy of Nature: From Systematic Patterns to Evolutionary Process Theories*, 1994. San Diego CA: Academic Press, pp. 7–31. Available online at: http://natureinstitute.org/txt/rb.

Brickell, J. S. and D. C. Wathes (2011). A Descriptive Study of the Survival of Holstein-Friesian Heifers through to Third Calving on English Dairy Farms, *Journal of Dairy Science* vol. 94(4), pp. 1831–8.

Britton, W. S. (1941). Form and Function in the Sloth, *Quarterly Review of Biology* vol. 16, pp.13–43 and 190-207.

Brooks, C. J. and Stephen Harris (2008). Directed Movement and Orientation Across a Large Natural Landscape by Zebras (*Equus burchelli antiquorum*), *Animal Behaviour* vol. 76, pp. 277–85. doi: 10.1016/j.anbehav.2008.02.005.

Brown, D. D. and L. Cai (2007). Amphibian Metamorphosis, *Developmental Biology* vol. 3016, pp. 20–33.

Brown, Joseph Epes (1989). *The Sacred Pipe: Black Elk's Account of the Seven Rites of the Oglala Sioux*. Norman, OK: University of Oklahoma Press.

——— (2007). *The Spiritual Legacy of the American Indian*. Bloomington, Indiana: World Wisdom Inc.

Brown, R. E. et al. (1997). The Elephant's Respiratory System Adaptations to Gravitational Stress, *Respiration Physiology* vol. 109, pp. 177–94.

Brownlee, A. (1963). Evolution of the Giraffe, *Nature* vol. 200, p. 1022.

Budiansky, Stephen (1997). *The Nature of Horses*. New York: The Free Press.

Buss, Irven O. (1990). *Elephant Life*. Ames, IA: Iowa State University Press.

Cabrera, Angel (1936). Subspecific and Individual Variation in the Burchell Zebras, *Journal of Mammalogy* vol. 17(2), pp. 89–112. doi: 10.2307/1374181.

Cameron, Elissa and Johan T. du Toit (2006). Winning by a Neck: Tall Giraffes Avoid Competing with Shorter Browsers, *The American Naturalist* vol. 169, pp. 130–5.

Caro, Tim (2016). *Zebra Stripes*. Chicago: University of Chicago Press.

Caro, Tim et al. (2014). The Function of Zebra Stripes, *Nature Communications* vol. 5, pp. 1–10. doi: 10.1038/ncomms4535.

Carrington, Richard (1959). *Elephants*. New York: Basic Books.

Carroll, Robert L. (1997). *Patterns and Processes of Vertebrate Evolution*. New York: Cambridge University Press.

Carroll, Robert L. (2000). The Lissamphibian Enigma. In: H. Heatwole and R. L. Carroll, Eds., *Amphibian Biology*. Chipping Norton, Australia: Surrey Beatty and Sons, pp. 1270–73.

Cartmill, F. and F. H. Smith (2009). *The Human Lineage*. Hoboken NJ: Wiley-Blackwell.

Castro-Vásquez, L. et al. (2010). Activity Patterns, Preference and Use of Floristic Resources by *Bradypus variegatus* in a Tropical Dry Forest Fragment, Santa Catalina, Bolivar, Colombia, *Edentata* vol. 11, pp. 62–9.

Catania, Kenneth, C. (2000). A Star is Born, *Natural History*, June, pp. 66–9.

——— (2002). The Nose Takes a Starring Role, *Scientific American*, July, pp. 54–9.

——— (2011). The Sense of Touch in the Star-Nosed Mole: From Mechanoreceptors to the Brain, *Philosophical Transactions of the Royal Society B* vol. 366, pp. 3016–25.

Catania, Kenneth, C. and Jon H. Kaas (1995). Organization of the Somatosensory Cortex of the Star-Nosed Mole, *The Journal of Comparative Neurology* vol. 351, pp. 549–67.

——— (1997). Somatosensory Fovea in the Star-Nosed Mole: Behavioral Use of the Star in Relation to Innervation Patterns and Cortical Representation, *The Journal of Comparative Neurology* vol. 387, pp. 215–33.

Cauvin, Jacques (2000). *The Birth of the Gods and the Origins of Agriculture*. Cambridge, UK: Cambridge University Press.

Chadwick, Douglas (1992). *The Fate of the Elephant*. San Francisco, CA: Sierra Club Books.

Chevalier-Skolnikoff, Suzanne and J. Liska (1993). Tool Use by Wild and Captive Elephants, *Animal Behavior* vol. 46, pp. 209–19.

Churcher, C. S. (1978). Giraffidae. In: Maglio, V. J. and H. B. S. Cooke, Eds., *Evolution of Mammals in Africa*. Princeton: Princeton University Press, pp. 509–35.

Ciofolo, I. and Y. Le Pendu (2002). The Feeding Behaviour of Giraffe in Niger, *Mammalia* vol. 66, pp. 183–94.

Clack, J. A. (2006). The Emergence of Early Tetrapods, *Palaeogeography, Palaeoclimatology, Palaeoecology* vol. 232, pp. 167-89.

Clack, J. A. (2012). *Gaining Ground* (second edition). Bloomington IN: University of Indiana Press.

Cliffe, Rebecca N. et al. (2015). Sloths Like it Hot: Ambient Temperature Modulates Food Intake in the Brown-throated Sloth (*Bradypus variegatus*), *PeerJ* 3:e875; doi: 10.7717/peerj.875.

Coe, Malcolm J. (1967). "Necking" Behavior in the Giraffe, *Journal of Zoology, London* vol. 151, pp. 313–21.

Colbert, Edwin H. (1935). The Classification and Phylogeny of the Giraffidae, *American Museum Novitates* no. 800, pp. 1–15.

Colbert, Edwin H. (1938). Relationships of the Okapi, *Journal of Mammalogy* vol.19, pp. 47–64.

Coppedge, Bryan R. (2009). Patterns of Bison Hair Use in Nests of Tallgrass Prairie Birds, *The Prairie Naturalist* vol. 41, pp. 110–5.

Coppedge, Bryan R. and James Shaw (2000). American Bison *Bison bison* Wallowing Behavior and Wallow Formation on Tallgrass Prairie, *Acta Theriologica* vol. 45, pp. 103–10.

Dagg, Ann Innis and J. Bristol Foster (1982). *The Giraffe: Its Biology, Behavior and Ecology.* Malabar, FL: Krieger Publishing Company.

Daniel, J. C. (1998). *The Asian Elephant—A Natural History,* Dehra DunI, India: Natraj Publishers.

Danowitz, Melinda et al. (2015a). Fossil Evidence and Stages of Elongation of the *Giraffa camelopardalis* Neck, *Royal Society Open Science* vol. 2, p. 150393.

Danowitz, Melinda et al. (2015b). The Cervical Anatomy of *Samotherium*, an Intermediate-necked Giraffid, *Royal Society Open Science* vol. 2, p. 150521.

Darwin, Charles (1872). *Origin of Species.* Sixth Edition. (Available online at: http://darwin-online.org.uk/content/frameset?itemID=F391&viewtype=side&pageseq=1.)

Dawkins, Richard (1996). *Climbing Mount Improbable.* New York: W. W. Norton & Company.

de Almeida, S. F. and M. Carmo-Fonseca (2012). Design Principles of Interconnections between Chromatin and pre-mRNA Splicing, *Trends in Biochemical Sciences* vol. 37, no 6 (June), pp. 248–53. doi:10.1016/j.tibs.2012.02.002.

de Blainville, M. H. M. D. (1840). *Ostéographie ou description iconographique comparée du squelette et du système dentaire des cinq classes d'animaux vertébrés récents et fossiles pour servir de base a la zoologie et a la géologie. Mammifères.—Paresseaux.—G. Bradypus.* Arthus Bertrand, Part 4 (fasciculus 5, atlas [fasciculus 4]). Paris: Arthus Bertrand.

DeMallie, Raymond J., Ed. (1985). *The Sixth Grandfather: Black Elk's Teachings Given to John G. Neihardt.* Lincoln: University of Nebraska Press.

Dembski, William A. and Jonathan Wells (2008). *The Design of Life.* Seattle: Discovery Institute Press.

Dobzhansky, T. (1973). Nothing in Biology Makes Sense Except in the Light of Evolution, *The American Biology Teacher* vol. 35, pp. 125–9.

Dodd, M. H. I. and Dodd, J. M. (1976). The Biology of Metamorphosis. In: B. Lofts, Ed., *Physiology of Amphibia* vol. III. New York: Academic Press, pp. 467–599.

Dodge, Richard Irving (1877). *The Hunting Grounds of the Great West.* London: Chatto & Windus, Piccadilly.

Douglas-Hamilton, Iain et al. (2006). Behavioural Reactions of Elephants Towards a Dying and Deceased Matriarch, *Applied Animal Behaviour Science* vol. 100, pp. 87–102.

Duellman, W. E. and L. Trueb (1994). *Biology of Amphibians.* Baltimore MD: The John Hopkins University Press.

Du Toit, J. T. (1990a). Feeding-height Stratification Among African Browsing Ruminants, *African Journal of Ecology* vol. 28, pp. 55–61.

———— (1990b). Giraffe Feeding on *Acacia* Flowers: Predation or Pollination? *African Journal of Ecology* vol. 28, pp. 63–8.

Ebach, Malte C. and David M. Williams (2019). Ronald Brady and the Cladists. *Cladistics* vol. 35. doi: 10.1111/cla.12397.

El Aich, A. et al. (2007). Ingestive Behavior of Goats Grazing in the Southwestern Argan (*Argania* spinosa) Forest of Morocco, *Small Ruminant Research* vol. 70, pp. 248–56.

Eicher, S. D. et al. (2006). Short Communication: Behavioral and Physiological Indicators of Sensitivity or Chronic Pain Following Tail Docking, *Journal of Dairy Science* vol. 89, pp. 3047–51.

Eisley, Loren (1961). *Darwin's Century.* Garden City, NY: Anchor Books.

Eldredge, N. and S. J. Gould (1972). Punctuated Equilibria: An Alternative to Phyletic Gradualism. In: Schopf, T., Ed. (1972). *Models in Paleobiology.* San Francisco, CA: Freeman, Cooper, pp. 82–115.

Elliott, John P. et al. (1977). Prey Capture by the African Lion, *Canadian Journal of Zoology* vol. 55(11), pp. 1811–28. doi: 10.1139/z77-235.

Emerson, Ralph Waldo (1983). *Nature* (Essays, Second Series) in Emerson, R. W. *Essays and Lectures.* New York: Library of America. (This essay was first published in 1844.)

Espejo, L. A. et al. (2006). Prevalence of Lameness in High-Producing Holstein Cows Housed in Freestall Barns in Minnesota, *Journal of Dairy Science* vol. 89, pp. 3052–8.

Estes, Richard D. (1991). *The Behavior Guide to African Mammals.* Berkeley, CA: University of California Press.

Evans, S. E. and M. Borsuk-Bialynick (2009). The Early Triassic Stem-Frog *Czatkobatrachus* from Poland, *Palaeontologica Polonica* vol. 65, pp. 79–105.

Fay, Philip A. (2003). Insect Diversity in Two Burned and Grazed Grasslands, *Environmental Entomology* vol. 32, pp. 1099–104.

Fennessy, J. T. et al. (2003). Distribution and Status of the Desert-dwelling Giraffe (*Giraffa camelopardalis angolensis*) in Northwestern Namibia, *African Zoology* vol. 38, pp. 184–8.

Fischhoff, Ilya R. et al. (2007). Habitat Use and Movements of Plains Zebra (*Equus burchelli*) in Response to Predation Danger from Lions, *Behavioral Ecology* vol. 18(4), pp. 725–9. doi: 10.1093/beheco/arm036.

Flindt, Rainer (1986). *Biologie in Zahlen.* Stuttgart: Gustav Fischer Verlag.

Flores, Dan (1991). Bison Ecology and Bison Diplomacy: The Southern Plains from 1800 to 1850, *The Journal of American History* vol. 78, pp. 465–85.

———— (2016). *American Serengeti: The Last Big Animals of the Great Plains.* Lawrence, KS: University Press of Kansas.

Foley, Charles (2002). *The Effects of Poaching on Elephant Social Systems*. Ph.D. dissertation, Princeton University.

Foley, W. J. et al. (1995). The Passage of Digesta, Particle Size, and *in vitro* Fermentation Rate in the Three-Toed Sloth *Bradypus tridactylus*, *Journal of Zoology* vol. 236, pp. 681–96.

Freese, Curtis H. et al. (2007). Second Chance for the Plains Bison, *Biological Conservation* vol. 136, pp. 175–84.

Fuhlendorf, Samuel D. et al. (2009). Pyric Herbivory: Rewilding Landscapes through the Recoupling of Fire and Grazing, *Conservation Biology* vol. 23, pp. 588–98.

Fulwider, W. K. et al. (2008). Survey of Dairy Management Practices on One Hundred Thirteen North Central and Northeastern United States Dairies, *Journal of Dairy Science* vol. 91(4), pp. 1686–92.

Funston, P. J. et al. (1998). Hunting by Male Lions: Ecological Influences and Socioecological Implications, *Animal Behaviour* vol. 56, pp. 1333–45. doi:10.1006/anbe.1998.0884.

Funston, P. J. et al. (2001). Factors Affecting the Hunting Success of Male and Female Lions in the Kruger National Park, *Journal of Zoology* vol. 253, pp. 419–31. doi: 10.1017/S0952836901000395.

Gambaryan, P. P. (1974). *How Mammals Run*. New York: John Wiley & Sons.

Gardner, Bruce L. (2002). *American Agriculture in the Twentieth Century: How It Flourished and What It Cost*. Cambridge, MA: Harvard University Press.

Garretson, Martin S. (1938). *The American Bison: The Story of its Extermination as a Wild Species and Its Restoration under Federal Protection*. New York, NY: New York Zoological Society.

Gates, C. Cormack et al. (2010). *American Bison: Status Survey and Conservation Guidelines*. Gland, Switzerland: International Union for Conservation of Nature and Natural Resources.

Gepts, Paul and Roberto Papa (2002). Evolution During Domestication, *Encyclopedia of Life Sciences*, pp. 1–7.

Gerlanc, Nicole M. and Glennis Kaufman (2003). Use of Bison Wallows by Anurans on Konza Prairie, *The American Midland Naturalist* vol. 150, pp. 158–68.

Gilbert, S. F. (2016). Ecological Developmental Biology: Interpreting Developmental Signs, *Biosemiotics* vol. 9. pp. 51–60. https://doi.org/10.1007/s12304-016-9257-4.

Gilmore, D. P. et.al. (2000). An Update on the Physiology of Two- and Three-toed Sloths, *Brazilian Journal of Medical and Biological Research* vol. 33, pp. 129–46.

Gilmore, D. P. et.al. (2001). Sloth Biology: An Update on Their Physiological Ecology, Behavior and Role as Vectors of Arthropods and Arboviruses, *Brazilian Journal of Medical and Biological Research* vol. 34, pp. 9–25.

Ginnett, Tim and Montague Demment (1997). Sex Differences in Giraffe Foraging Behavior at Two Spatial Scales, *Oecologia* vol. 110, pp. 291–300.

Gobush, K. S. et al. (2008). Long-term Impacts of Poaching on Relatedness, Stress Physiology, and Reproductive Output of Adult Female African Elephants, *Conservation Biology* vol. 22.6, pp. 1590–9.

Goethe, J. W. von (1851). *Briefwechsel zwischen Goethe und Knebel (1774–1832), erster Theil*. Leipzig: Brockhaus.

———— (1995). *Scientific Studies*. Princeton: Princeton University Press.

———— (2002) *Werke Band XIII Naturwissenschaftliche Schriften*. Munich: Verlag C. H. Beck.

Goetz, Robert H. and E. N. Keen (1957). Some Aspects of the Cardiovascular System in the Giraffe, *Angiology* vol. 8, pp. 542–64.

Goffart, M. (1971). *Form and Function in the Sloth*. Oxford and New York: Pergamon Press.

Goldstein, Kurt (1995). *The Organism*. New York: Zone Books.

Goodwin, Brian (2001). *How the Leopard Changed its Spots: The Evolution of Complexity*. Princeton: Princeton University Press.

Goodwin, Brian, and Gerry Webster (1996). *Form and Transformation: Generative and Relational Principles in Biology*. Cambridge, UK: Cambridge University Press.

Gottardo, F. et al. (2011). The Dehorning of Dairy Calves: Practices and Opinions of 639 Farmers, *Journal of Dairy Science* vol. 94, pp. 5724–34.

Gould, Stephan Jay (1996). The Tallest Tale, *Natural History* vol. 105, pp. 18–26.

———— (2002). *The Structure of Evolutionary Theory*. Cambridge, MA: Harvard University Press.

Gould, Stephen Jay and Niles Eldrege (1993). Punctuated Equilibrium Comes of Age, *Nature* vol. 366, pp. 223–7.

Grand, Theodore et al. (1998). Structure of the Proboscis and Rays of the Star-Nosed Mole, *Condylura cristata*, *Journal of Mammalogy* vol. 79(2). pp. 492–501.

Gray, James (1968). *Animal Locomotion*. New York: W. W. Norton & Company.

Grassé, Pierre–P. (1955). *Traité de Zoologie, vol. XVII (Mammifères)*. Paris: Libraires de L'Académie de Médecine.

Grinnell, Georg Bird (1962). *Blackfoot Lodge Tales*. Lincoln: University of Nebraska Press.

Groves, Colin P. and C. Bell (2004). New Investigations on the Taxonomy of the Zebras genus *Equus*, subgenus *Hippotigris, Mammalian Biology* vol. 69(3), pp. 182–96. doi: 10.1078/1616-5047-00133.

Grzimek, Bernhard (1975). *Grzimek's Animal Life Encyclopedia* vol. 11 (Mammals II). New York: Van Nostrand Reinhold Company.

Gutmann, Wolfgang (1995). *Die Evolution hydraulischer Konstruktionen: Organismische Wandlung statt altdarwinistischer Anpassung*. Stuttgart: E. Schweizerbart.

Haas, Sarah K. et al. (2005). *Panthera leo, Mammalian Species* vol. 762, pp. 1–11. doi:10.1644/1545-1410(2005)762[0001:PL]2.0.CO;2.

Hadley, G. L. et al. (2006). Dairy Cattle Culling Patterns, Explanations, and Implications, *Journal of Dairy Science* vol. 89(6), pp. 2286–96.

Hahn, Eduard (1896). *Die Haustiere und ihre Beziehungen zur Wirtschaft des Menschen.* Leipzig: Verlag von Duncker & Humblot.

Hanks, John (1979). *The Struggle for Survival: The Elephant Problem.* New York: Mayflower Books.

Hansen, L. B. (2000). Consequences of Selection for Milk Yield from a Geneticist's Viewpoint, *Journal of Dairy Science* vol. 83, pp. 1145–50.

Hare, E. et al. (2006). Survival Rates and Productive Herd Life of Dairy Cattle in the United States, *Journal of Dairy Science* vol. 89(9), pp. 3713–20.

Hargens, Alan R. et al. (1987). Gravitational Haemodynamics and Oedema Prevention in the Giraffe, *Nature* vol. 329, pp. 59–60.

Harlan, Jack R. (1998). *The Living Fields: Our Agricultural Heritage.* Cambridge, UK: Cambridge University Press.

Harris, J. M. (1976). Pleistocene Giraffidae (Mammalia, Artiodactyla) from East Rudolf, Kenya. In: Savage, R. J. G. and Shirley C. Coryndon, Eds., *Fossil Vertebrates of South Africa.* London: Academic Press, pp. 283–332.

Hart, John and Teresa Hart (1989). Ranging and Feeding Behavior of Okapi (*Okapi johnstoni*) in the Ituri Forest of Zaire: Food Limitation in a Rain-Forest Herbivore? *Symposium of the zoological society of London* vol. 61, pp. 31–50.

Haug, Herbert (1970). *Der Makroskopische Aufbau des Grosshirns.* Berlin: Springer-Verlag.

———— (1987). Brain Sizes, Surfaces and Neuronal Sizes of the Cortex Cerebri: A Sterological Investigation of Man and his Variability and a Comparison with some Mammals (primates, whales, marsupials, insectivores, and one elephant), *The American Journal of Anatomy* vol. 180, pp. 126–42.

Hautier, Lionel et al. (2016). The Hidden Teeth of Sloths: Evolutionary Vestiges and the Development of a Simplified Dentition. *Scientific Reports* vol. 6, 277763; doi: 10.1038/srep27763.

Hayssen, Virginia (2010). *Bradypus variegatus, Mammalian Species* vol. 42, pp. 19-32.

Hayward, Matt W. and G. Kerley (2005). Prey Preferences of the Lion (*Panthera leo*), *Journal of Zoology* vol. 267, pp. 309–22. doi: 10.1017/S0952836905007508.

Hayward, Matt W. and R. Slotow (2009). Temporal Partitioning of Activity in Large African Carnivores: Tests of Multiple Hypotheses, *South African Journal of Wildlife Research* vol. 39, pp. 109–25. doi: 10.3957/056.039.0207.

Herder, Johann Gottfried (1982). *Herder's Werke, Vierter Band: Ideen zur Philosophie der Geschichte der Menschheit* [Ideas Concerning a Philosophy of the History of Humanity]. Berlin: Aufbau Verlag. (This book was first published in 1791.)

Heuer, Holger et al. (2011). Antibiotic Resistance Gene Spread due to Manure Application on Agricultural Fields, *Current Opinion in Microbiology* vol. 14, pp. 236–43.

Hildebrand, Milton (1974). *Analysis of Vertebrate Structure*. New York: John Wiley & Sons.

Ho, Mae-Wan and P. T. Saunders (1984). *Beyond Neo-Darwinism: An Introduction to the New Evolutionary Paradigm*. London: Academic Press.

Hofmann, R. R. (1973). *The Ruminant Stomach. Stomach Structure and Feeding Habits of East African Game Ruminants* vol. 2. Nairobi: East African Literature Bureau.

Hofmann, R. R. and B. Matern (1988). Changes in Gastrointestinal Morphology Related to Nutrition in Giraffes, *International Zoo Yearbook* vol. 27, pp. 168–76.

Holdrege, Craig (2005a). Doing Goethean Science, *Janus Head* vol. 8.1, pp. 27–52. Available online: http://www.janushead.org/8-1/holdrege.pdf.

——— (2005b). *The Giraffe's Long Neck: From Evolutionary Fable to Whole Organism*. Ghent, NY: The Nature Institute.

——— (2009). Evolution Evolving, *In Context* #21, pp. 16–23. Available online: http://www.natureinstitute.org/pub/ic/ic21/darwin.htm.

——— (2013). *Thinking Like a Plant: A Living Science for Life*. Great Barrington, MA.: Lindisfarne Books.

——— (2014). Goethe and the Evolution of Science, *In Context* #31 pp. 10–23. Available online: http://natureinstitute.org/pub/ic/ic31/goethe.pdf.

——— (2016). Meeting Nature as a Presence: Aldo Leopold and the Deeper Nature of Nature, *In Context* #36, pp. 14–17.

Holechek, Jerry L. (1984). Comparative Contribution of Grasses, Forbs, and Shrubs to the Nutrition of Range Ungulates, *Rangelands* vol. 6(6), pp. 261–3.

Hornaday, William T. (1889/2002). *The Extermination of the American Bison*. Washington DC: Smithsonian Institution Press.

Horváth, Gábor et al. (2018). Experimental Evidence That Stripes Do Not Cool Zebras, *Scientific Reports* vol. 8:935; doi: 10.1038/s41598-018-27637-1.

Humane Society of the United States (2012). Welfare Issues with Tail Docking of Cows in the Dairy Industry, *HSUS Report*, October, pp. 1–8.

Hutchinson, John R. et al. (2003). Are Fast-moving Elephants Really Running? *Nature* vol. 422, pp. 493–4.

Huxley, Thomas Henry (1860). The Origin of Species, *Collected Essays* vol. II. Available online: http://aleph0.clarku.edu/huxley/CE2/OrS.html.

Irrgang, Nora (2012). *Horns in Cattle—Implications of Keeping Horned Cattle or Not*. Ph.D. Thesis, Universität Kassel, Witzenhausen.

Irwin, Lee (1996). *The Dream Seekers: Native American Visionary Traditions of the Great Plains*. Norman, OK: University of Oklahoma Press.

Isenberg, Andrew C. (2000). *The Destruction of the Bison: An Environmental History, 1750–1920*. Cambridge, UK: Cambridge University Press.

Isaac, Erich (1962). On the Domestication of Cattle, *Science* vol. 137, pp. 195–204.

Jablonka, Eva and M. Lamb (1995). *Epigenetic Inheritance and Evolution*. Oxford: Oxford University Press.

Jablonka, Eva and M. Lamb (2005). *Evolution in Four Dimensions*. Cambridge, MA: The MIT Press.

Jablonka, Eva and Gal Raz (2009). Transgenerational epigenetic inheritance: prevalence, mechanisms, and implications for the study of heredity and evolution. *The Quarterly Review of Biology* vol. 84, pp. 131–76.

Jami, Elie et al. (2013). Exploring the Bovine Rumen Bacterial Community from Birth to Adulthood, *The ISME Journal* vol. 7, pp. 1069–79.

Jonas, Hans (1966). *The Phenomenon of Life*. New York: Harper and Row.

———— (1968). Biological Foundations of Individuality, *International Philosophical Quarterly* vol. 8, pp. 231–51.

Jung, Thomas S. et al. (2010). Bison (*Bison bison*) Fur used as Drey Material by Red Squirrels (*Tamiasciurus hudsonicus*): An Indication of Ecological Restoration, *Northwestern Naturalist* vol. 91(2), pp. 220–2.

Kant, Immanuel (1951). *Critique of Judgement*. Translated by J. H. Bernard. New York: Hafner Press. (Originally published in 1790 in German: *Kritik der Urteilskraft*.)

Kehoe, Thomas F. (1965). Buffalo Stones: An Addendum to The Folklore of Fossils, *Antiquity* vol. 39, pp. 212–3.

Keller, Evelyn Fox (2008). Organisms, Machines, and Thunderstorms: A History of Self-Organization, Part One, *Historical Studies in the Natural Sciences* vol. 38.1 pp. 45–75.

Kilgannon, Corey (2018). When the Death of Family Farm Leads to Suicide, *New York Times* (March 19). https://www.nytimes.com/2018/03/19/nyregion/farmer-suicides-mark-tough-times-for-new-york-dairy-industry.html.

Kimani, J. K. and J. M. Mungai (1983). Observations on the Structure and Innervation of the Presumptive Carotid Sinus Area in the Giraffe (*Giraffa camelopardalis*), *Acta Anat* (Basel) vol. 115, pp. 117–33.

Kimani, J. K. and I. O. Opole (1991). The Structural Organization and Adrenergic Innervation of the Carotid Arterial System of the Giraffe (*Giraffa camelopardalis*), *The Anatomical Record* vol. 230, pp. 369–77.

Kingdon, Jonathan (1977). *East African Mammals, Volume IIIA*. Chicago: University of Chicago Press.

———— (1989). *East African Mammals, Volume IIIB, Large Mammals*. Chicago: University of Chicago Press.

Klingel, Hans (1969). The Social Organisation and Population Ecology of the Plains Zebra (*Equus quagga*), *Zoologica Africana* vol. 4, pp. 249–63. doi: 10.1080/00445 096.1969.11447374.

Klingel, Hans and Uta Klingel (1966). Die Geburt eines Zebras (*Equus quagga böhmi Matschie*), *Zeitschrift für Tierpsychologie* vol. 23, pp. 72–6.

Knapp, Alan K. et al. (1999). The Keystone Role of Bison in North American Tallgrass Prairie: Bison Increase Habitat Heterogeneity and Alter a Broad Array of Plant, Community, and Ecosystem Processes, *BioScience* vol. 49, pp. 39–50.

Knauer, Vinzenz. (1892). *Die Hauptprobleme der Philosophie*. Vienna: Wilhelm Braunmüller.

Knaus, Wilhelm (2009). Dairy Cows Trapped between Performance Demands and Adaptability, *Journal of the Science of Food and Agriculture* vol. 89(7), pp. 1107–14.

Knierim, Ute et al. (2015). To Be or Not To Be Horned—Consequences in Cattle, *Livestock Science* vol. 179.

König, H. E. (1983). Osteologie des Giraffenschädels, *Tierärtzl. Prax.* vol. 11, pp. 405–15.

Krakauer, Alan H. and T. H. Krakauer (1999). Foraging of Yellow-headed Caracaras in the Fur of a Three-toed Sloth, *Journal of Raptor Research* vol 33, p. 270.

Kranich, Ernst-Michael (1995). *Wesensbilder der Tiere*. Stuttgart: Verlag Freies-Geistesleben.

———— (1999). *Thinking Beyond Darwin: The Idea of the Type as a Key to Vertebrate Evolution*. Great Barrington, MA: Lindisfarne Books.

Krause, K. Marie and Garrett Oetzel (2006). Understanding and Preventing Subacute Ruminal Acidosis in Dairy Herds: A Review, *Animal Feed Science and Technology* vol. 126, pp. 215–36.

Krumbiegel, Ingo (1971). *Die Giraffe*. Wittenberg: A. Ziemsen Verlag.

Lamarck, J. B. (1984). *Zoological Philosophy*. Chicago: University of Chicago Press.

Lampi, M. et al. (1992). Saltation and Stasis: A Model of Human Growth, *Science* vol. 258, p. 801–3.

Langer, Susanne (1967). *Mind: An Essay on Human Feeling*. Baltimore: The John Hopkins Press.

Langman, Vaughan A. (1982). Giraffe Youngsters Need a Little Bit of Maternal Love, *Smithsonian* vol. 12, pp. 94–103.

Langman, V. A. et al. (1982). Respiration and Metabolism in the Giraffe, *Respiration Physiology* vol. 50, pp. 141–52.

Lankester, Ray (1908). On Certain Points in the Structure of the Cervical Vertebrae of the Okapi and the Giraffe, *Journal of Zoology* vol. 78, pp. 320–34.

Larison, Brenda et al. (2015a). How the Zebra Got its Stripes: A Problem with Too Many Solutions, *Royal Society Open Science* vol. 2. doi: 10.1098/rsos.140452.

Larison, Brenda et al. (2015b). Concordance on Zebra Stripes is not Black and White: Response to Comment by Caro & Stankowich (2015), *Royal Society Open Science* vol. 2. doi: 10.1098/rsos.150359.

Laws, R. M. (1966). Age Criteria for the African Elephant, *Loxodonta a. Africana*, *East African Wildlife Journal* vol. 4, pp. 1–37.

Ledón-Rettig, C. C. and Pfennig, D. W. (2011). Emerging Model Systems in Eco-Evo-Devo: The Environmentally Responsive Spadefoot Toad, *Evolution and Development* vol. 13, pp. 391–400. doi:10:1111/j.1525-142X.2011.00494.x.

Leopold, Aldo (1987). *A Sand County Almanac*. Oxford: Oxford University Press.

Leuthold, Barbara and Walter Leuthold (1972). Food Habits of Giraffe in Tsavo National Park, Kenya, *African Journal of Ecology* vol. 10, pp. 129–41.

——— (1978). Ecology of the Giraffe in Tsavo National Park, Kenya, *African Journal of Ecology* vol. 16, pp. 1–20.

Levit, Georgy S. et al. (2008). Alternative Evolutionary Theories: A Historical Survey, *Journal of Bioeconomics* vol. 10, pp. 71–96. doi: 10.1007/s10818-008-9032-y.

Lindberg, Ann Morrow (1966). Immersion in Life: Journey to East Africa, *Life* vol 61(17), pp. 89–99.

Linderman, Frank Bird (2002). *Plenty-coups, Chief of the Crows*. Lincoln: University of Nebraska Press.

Lindsey, Susan L. et al. (1999). *The Okapi*. Austin: University of Texas Press.

Lockley, Martin (1999). *The Eternal Trail: A Tracker Looks at Evolution*. Reading, MA: Perseus Books.

——— (2007). The Morphodynamics of Dinosaurs, Other Archosaurs, and their Trackways: Holistic Insights into Relationships between Feet, Limbs, and the Whole Body, *Special Publication-SEPM* vol. 88, pp. 27–51.

Lordkipanidze, D. et al. (2013). A Complete Skull from Dmanisi, Georgia, and the Evolutionary Biology of Early *Homo, Science v*ol. 342. pp. 326–31. *doi:*10.1126/science.1238484.

Lott, Dale F. (2002). *American Bison: A Natural History*. Berkeley, CA: University of California Press.

Lovejoy, C. O. et al. (2009). The Great Divides: *Ardipithecus ramidus* Reveals the Postcrania of Our Last Common Ancestor with African Apes, *Science* vol. 32, pp. 100–6. doi: 10.1126/science.1175833.

Lucy, M. C. (2001). Reproductive Loss in High-Producing Dairy Cattle: Where Will It End? *Journal of Dairy Science* vol. 84(6), pp. 1277–93.

Madden, Derek and Truman P. Young (1992). Symbiotic Ants as an Alternative Defense Against Giraffe Herbivory in Spinescent *Acacia drepanolobium*, *Oecologia* vol. 91, pp. 235–8.

Marshall, Bonnie. M. and Stuart B. Levy (2011). Food Animals and Antimicrobials: Impacts on Human Health, *Clinical Microbiology Reviews* vol. 24, pp. 718–33. doi:10.1128/CMR.00002-11.

McComb, Karen et al. (2001). Matriarchs as Repositories of Social Knowledge in African Elephants, *Science* vol. 292, pp. 491–4.

McComb, Karen et al. (2005). African Elephants Show High Levels of Interest in the Skulls and Ivory of their own Species, *Biology Letters* vol. 2, pp. 26–8.

McDiarmid, R. W. and R. Altig, Eds. (1999). *Tadpoles*. Chicago: The University of Chicago Press.

McDonald, Jerry N. (1981). *North American Bison: Their Classification and Evolution*. Berkeley, CA: University of California Press.

McFadden, Brian (1999). *Fossil Horses*. Cambridge, UK: Cambridge University Press.

McHugh, Tom (1979). *The Time of the Buffalo*. Lincoln, NE: University of Nebraska Press.

McIntosh, Wendy et al. (2016). Suicide Rates by Occupational Group – 17 States, 2012, *Centers for Disease Control and Prevention, Morbidity and Mortality Weekly Report* vol. 65, no. 25, pp. 641–5.

McKenna, Maryn (2107). *Big Chicken*. Washington DC: National Geographic Partners.

McMillan, Brock R. et al. (2011). Vegetation Responses to an Animal-generated Disturbance (Bison Wallows) in Tallgrass Prairie, *The American Midland Naturalist* vol. 165, pp. 60–73.

McNab, Brian K. (1978). Energetics of Arboreal Folivores: Physiological Problems and Ecological Consequences of Feeding on an Ubiquitous Food Supply. In: G. Gene Montgomery, Ed, *The Ecology of Arboreal Folivores*. Washington D.C.: Smithsonian Institution Press.

McNaughton, S. J. (1983). Compensatory Plant Growth as a Response to Herbivory, *Oikos* vol. 40, pp. 329–36.

———— (1985). Ecology of a Grazing Ecosystem: The Serengeti, *Ecological Monographs* vol. 55(3), pp. 259–94. doi: 10.2307/1942578.

McNaughton, S. J. et al. (1985). Silica as a Defense against Herbivory and a Growth Promotor in African Grasses, *Ecology* vol. 66(2), pp. 528–35. doi: 10.2307/1940401.

Melin, Amanda D. et al. (2016). Zebra Stripes through the Eyes of Their Predators, Zebras, and Humans, *PLOS ONE* vol. 11(3). doi: 10.1371/journal.pone.0151660.

Mendel, Frank C. (1985a). Adaptations for Suspensory Behavior in the Limbs of Two-toed Sloths. In: G. Gene Montgomery (Ed), *The Evolution and Ecology of Armadillos, Sloths, and Vermilinguas*. Washington D.C.: Smithsonian Institution Press.

———— (1985b). Use of Hands and Feet of Three-toed Sloths (*Bradypus variegatus*) During Climbing and Terrestrial Locomotion, *Journal of Mammalogy* vol. 66, pp. 359–66.

Mendel, Frank C. et al. (1985). Vision of Two-toed Sloths (*Choloepus*), *Journal of Mammalogy* vol. 66, pp. 197–200.

Meyer, L. M. et al. (2013). How to Understand the Gene in the Twenty-First Century? *Science and Education* vol. 22, pp. 345–74. doi:10.1007/s11191-011-9390-z.

Meyer, Stephen (2004). The Origin of Biological Information and the Higher Taxonomic Categories, *Proceedings of the Biological Society of Washington* vol. 117, pp. 213–39.

Milewski, A. V. et al. (1991). Thorns as Induced Defenses: Experimental Evidence, *Oecologia* vol. 86, pp. 70–5.

Milius, Susan (2003). Beast Buddies: Do Animals Have Friends? *Science News* vol. 164, pp. 282–4.

Mitchell, G. and J. D. Skinner (1993). How Giraffe Adapt to Their Extraordinary Shape, *Transactions of the Royal Society of South Africa* vol. 48, pp. 207–18.

———— (2003). On the Origin, Evolution and Phylogeny of Giraffes (*Giraffa Camelopardalis*), *Transactions of the Royal Society of South Africa* vol. 58, pp. 51–73.

———— (2009). An Allometric Analysis of the Giraffe Cardiovascular System, *Comparative Biochemistry and Physiology, Part A* vol. 154, pp. 523–9.

Mitchell, G. et al. (2010). The Demography of Giraffe Deaths in a Drought, *Transactions of the Royal Society of South Africa* vol. 65, pp. 165–8.

Mitchell, G. et al. (2013). Orbit Orientation and Eye Morphometrics in Giraffes (*Giraffa camelopardalis*), *African Zoology* vol. 48, pp. 333–9.

Mitchell, G. et al. (2017). Body Surface Area and Thermoregulation in Giraffes, *Journal of Arid Environments* vol. 145, pp. 35–42.

Montgomery, G. Gene and M. E. Sunquist (1975). Impact of Sloths on Neotropical Forest Energy Flow and Nutrient Cycling. In: Frank. B. Golley and Ernesto Medina (Ed), *Tropical Ecological Systems*. New York: Springer Verlag.

———— (1978). Habitat Selection and Use by Two-toed and Three-toed Sloths. In: G. Gene Montgomery, Ed., *The Ecology of Arboreal Folivores*. Washington D.C.: Smithsonian Institution Press, pp. 329-59.

Moss, Cynthia (1982). *Portraits in the Wild* (2nd edition). Chicago: University of Chicago Press.

———— (1988). *Elephant Memories*. New York: William Morrow and Company, Inc.

Moss, Cynthia and Martyn Colbeck (1992). *Echo of the Elephants*. New York: William Morrow and Company, Inc.

Moss, Cynthia et al. (2011). *The Amboseli Elephants*. Chicago: University of Chicago Press.

Mosser, Anna and Craig Packer (2009). Group Territoriality and the Benefits of Sociality in the African Lion, *Panthera leo, Animal Behavior* vol. 78, pp. 359–70.

Muir, John. (2003). *My First Summer in the Sierra*. New York: The Modern Library.

Murie, James (1872). On the Horns, Viscera, and Muscles of the Giraffe, *Annals and Magazine of Natural History* vol. 9 (no. 51), pp. 177–95.

Nagel, Thomas (1974). What is it Like to be a Bat? *The Philosophical Review* vol. 83 (4), pp. 435–50.

Naidoo, R. et al. (2016). A Newly Discovered Wildlife Migration in Namibia and Botswana Is the Longest in Africa, *Oryx* vol. 50, pp. 138–46. doi: 10.1017/S003060531400022.

Naples, Virginia L. (1982). Cranial Osteology and Function in the Tree Sloths, *Bradypus* and *Choloepus*, *American Museum Novitates* vol. 2739, pp. 1–41.

———— (1985). The Superficial Facial Musculature in Sloths and Vermilinguas (anteaters). In: G. Gene Montgomery (Ed), *The Evolution and Ecology of Armadillos, Sloths, and Vermilinguas*. Washington D.C.: Smithsonian Institution Press, pp. 173–89.

National Research Council (2001). *Nutrient Requirements of Dairy Cattle*. Washington, DC: National Academy Press.

Nicholson, Daniel J. and John Dupré, eds. (2018). *Everything Flows: Towards a Processual Philosophy of Biology*. Oxford: Oxford University Press.

Nickel, Richard et al. (1986). *The Anatomy of the Domestic Animals. Vol. 1, The Locomotor System of the Domestic Mammals*. Berlin: Verlag Paul Parey.

O'Connell-Rodwell, Caitlin E. (2007). Keeping an "Ear" to the Ground: Seismic Communication in Elephants, *Physiology* vol. 22, pp. 287–94.

O'Connell-Rodwell, C. E. and B. T. Arnason (2000). Seismic Properties of Asian Elephant (*Elephas maximus*) Vocalizations and Locomotion, *The Journal of the Acoustical Society of America* vol. 108.6, pp. 3066–72.

Oltenacu, P. A. and D. M. Broom (2010). The Impact of Genetic Selection for Increased Milk Yield on the Welfare of Dairy Cows, *Animal Welfare* vol. 19(1), pp. 39–49.

Østergaard, Kristine et al. (2013). Left Ventricular Morphology of the Giraffe Heart Examined by Stereological Methods, *The Anatomical Record* vol. 296, pp. 611–21.

Owen, Richard (1839). Notes on the Anatomy of the Nubian Giraffe, *Journal of Zoology* vol. 2, pp. 217–43.

———— (1868/2011). *On the Anatomy of Vertebrates, Vol. 3, Mammals*. Cambridge, UK: Cambridge University Press.

Owen-Smith, Norman and J. Chafota (2012). Selective Feeding by a Megaherbivore, the African Elephant (*Loxodonta Africana*), *Journal of Mammalogy* vol. 93, pp. 698–705.

Pauli, Jonathan N. et al. (2014). A Syndrome of Mutualism Reinforces the Lifestyle of a Sloth, *Proceedings of the Royal Society of London B: Biological Sciences* vol. 281, pp. 1–7.

Payne, Katharine (1989). Elephant Talk, *National Geographic* vol. 176, pp. 264–77.

———— (1998). *Silent Thunder*. New York: Simon & Schuster.

Payne, Katharine et al. (1986). Infrasonic Calls of the Asian Elephant (*Elephas maximus*), *Behavioral Ecology and Sociobiology* vol. 18, pp. 297–301.

Pellew, Robin (1984). The Feeding Ecology of a Selective Browser, the Giraffe (*Giraffa camelopardalis tippelskirchi*), *Journal of Zoology* vol. 202, pp. 57–81.

Pfefferle, Dana et al. (2007). Do Acoustic Features of Lion, *Panthera leo,* Roars Reflect Sex and Male Condition? *The Journal of the Acoustical Society of America* vol. 121, pp. 3947–53.

Pfennig, D. W. (1992). Polyphenism in Spadefoot Toad Tadpoles as a Locally Adjusted Evolutionarily Stable Strategy, *Evolution* vol. 46, pp. 1408–20.

Piggins, David and W. R. A. Muntz (1985). The Eye of the Three-toed Sloth. In: G. Gene Montgomery (Ed), *The Evolution and Ecology of Armadillos, Sloths, and Vermilinguas*. Washington D.C.: Smithsonian Institution Press.

Pincher, Chapman (1949). Evolution of the Giraffe, *Nature* vol. 164, pp. 29–30.

Plaizier, J. C. et al. (2008). Subacute Ruminal Acidosis in Dairy Cows: The Physiological Causes, Incidence and Consequences, *The Veterinary Journal* vol. 176.1, pp. 21–31.

Poole, Joyce (1996). *Coming of Age with Elephants*. New York: Hyperion.

Poole, Joyce et al. (2005). Elephants are Capable of Vocal Learning, *Nature* vol. 434, pp. 455–6.

Portmann, Adolf (1967). *Animal Forms and Patterns*. New York: Schocken Books.

Posthumus, David C. (2016). A Lakota View of Pté Oyáte (Buffalo Nation). In: Geoff Cunfer and Bill Waiser, Eds., *Bison and People on the North American Great Plains: A Deep Environmental History*. College Station: Texas A&M University Press, pp. 278–309.

Pratt, David and Virginia Anderson (1979). Giraffe Cow-Calf Relationships and Social Development of the Calf in the Serengeti, *Ethology* vol. 51, pp. 233–51.

Price, Edward O. (1984). Behavioral Aspects of Animal Domestication, *The Quarterly Review of Biology* vol. 59(1), pp. 1–32.

——— (1999). Behavioral Development in Animals Undergoing Domestication, *Applied Animal Behaviour Science* vol. 65(3), pp. 245–71.

Probst, J. K. (2017). Unterscheiden sich die Schädelformen von behornten und unbehornten Kühen? Poster at: 14. Wissenschaftstagung Ökologischer Landbau, Campus Weihenstephan, Freising-Weihenstephan, 07.-10. März 2017. http://orgprints.org/31723/.

Pyne, Stephen J. (1988). *Fire in America: A Cultural History of Wildland and Rural Fire*. Princeton, NJ: Princeton University Press.

Ramsay, Edward C. and Robert W. Henry (2001). Anatomy of the Elephant Foot. In: Blair Csuti et al., Eds. *The Elephant's Foot*. IA: Iowa State University Press.

Rasmussen, L. E. L. and Bryce L. Munger (1996). The Sensorineural Specializations of the Trunk Tip (Finger) of the Asian Elephant, *Elephas maximus*, *The Anatomical Record* vol. 246, pp. 127–34.

Rasmussen, L. E. L. et al. (2002). Mellifluous Matures to Malodorous in Musth, *Nature* vol. 415, pp. 975–6.

Rees, P. A. (1982). Gross Assimilation Efficiency and Food Passage Time in the African Elephant, *African Journal of Ecology* vol. 20, pp. 193–8.

Reuter, Tom et al. (1998). Elephant Hearing, *Journal of the Acoustical Society of America* vol. 104, pp. 1122–3.

Richards, Robert J. (2002). *The Romantic Conception of Life: Science and Philosophy in the Age of Goethe*. Chicago: University of Chicago Press.

Riedel, Rupert (1978). *Order in Living Organisms*. Chichester and New York: John Wiley & Sons.

Riegner, Mark (1998). Horns, Hooves, Spots, and Stripes: Form and Pattern in Mammals. In: Seamon, David and Arthur Zajonc, Eds., *Goethe's Way of Science*. Albany, NY: SUNY Press, pp. 177–212.

———— (2008). Parallel Evolution of Plumage Pattern and Coloration in Birds: Implications for Defining Avian Morphospace, *The Condor* vol. 110, pp. 599–614.

Roček, Z. (2000). Mesozoic Anurans. In: H. Heatwole and R. L. Carroll, Eds., *Amphibian Biology*, Chipping Norton, Australia: Surrey Beatty and Sons, pp. 1295–331.

Roček, Z. and J.-C. Rage (2000). Proanuran Stages. In: H. Heatwole and R. L. Carroll, Eds., *Amphibian Biology*, Chipping Norton, Australia: Surrey Beatty and Sons, pp. 1283–94

Roček, Z. and E. van Dijk (2006). Patterns of Larval Development in Cretaceous Pipid Frogs, *Acta Palaeontologica Polonica* vol. 51, pp. 111–26.

Roček, Z. et al. (2012). Post-Metamorphic Development of Early Cretaceous Frogs as a Tool for Taxonomic Comparisons, *Journal of Vertebrate Paleontology* vol. 32, pp. 1285–92.

Romer, Alfred (1966). *Vertebrate Paleontology*. Chicago: University of Chicago Press.

Rosas, Claudia A. et al. (2008). Seed Dispersal by *Bison bison* in a Tallgrass Prairie, *Journal of Vegetation Science* vol. 19, pp. 769–78.

Rosslenbroich, Bernd (2014). *On the Origin of Autonomy: A New Look at Major Transitions in Evolution*. Heidelberg/New York: Springer Verlag.

Russel, James and Jennifer Rychlik (2001). Factors That Alter Rumen Microbial Ecology, *Science* vol. 292, pp. 1119–22.

Russell, Nerissa (2011). *Social Zooarchaeology: Humans and Animals in Prehistory*. New York: Cambridge University Press.

Ruxton, Graeme. D. (2002). The Possible Fitness Benefits of Striped Coat Coloration for Zebra, *Mammal Review* vol. 32, pp. 237–44. doi:10.1046/j.1365-2907.2002.00108.x.

Sachs, Joe (2005). Aristotle: Motion and its Place in Nature, *Internet Encyclopedia of Philosophy*. http://www.iep.utm.edu/aris-mot/.

Sanderson, Eric W. et al. (2008). The Ecological Future of the North American Bison: Conceiving Long-Term, Large-Scale Conservation of Wildlife, *Conservation Biology* vol. 22, pp. 252–66.

Sanson, Gordon (2006). The Biomechanics of Browsing and Grazing, *American Journal of Botany* vol. 93(10), pp. 1531–45.

Sanson, Gordon D. et al. (2007). Do Silica Phytoliths Really Wear Mammalian Teeth? *Journal of Archaeological Science* vol. 34(4), pp. 526–31. doi: 10.1016/j. jas.2006.06.009.

Sayers, K. et al. (2012). Human Evolution and the Chimpanzee Referential Doctrine, *Annual Review of Anthropology* vol. 41, pp. 119–38. doi: 10.1146/annurev-anthro-092611-145815

Schad, Wolfgang. (1977). *Man and Mammals*. Garden City, NY: Waldorf Press.

Schad, Wolfgang (2018). *Understanding Mammals: Threefoldness and Diversity*. Ghent, NY: Adonis Press.

Schaller, George (1972). *The Serengeti Lion*. Chicago: University of Chicago Press.

Schindewolf, Otto H. (1950/1993). *Basic Questions in Paleontology*. Chicago: University of Chicago Press.

——— (1969). *Über den "Typus" in morphologischer und phylogenetischer Biologie*. Mainz: Akademie der Wissenschaften und der Literatur.

Schoch, Rainer R. (2009). Evolution of Life Cycles in Early Amphibians, *Annual Review of Earth and Planetary Sciences* vol. 37, pp. 135–62.

Schultz, A. H. (1930). The Skeleton of the Trunk and Limbs of Higher Primates, *Human Biology* vol. 2, pp. 303–438.

——— (1936). Characters Common to Higher Primates and Characters Specific for Man, *The Quarterly Review of Biology* vol. 11, pp. 259–83.

Schultz-Bergin, Marcus (2017). The Dignity of Diminished Animals: Species Norms and Engineering to Improve Welfare. *Ethical Theory and Moral Practice* vol. 20, pp. 843–56.

Seamon, David and Arthur Zajonc, Eds. (1998). *Goethe's Way of Science*. Albany, NY: SUNY Press.

Shaw, James H. (1995). How Many Bison Originally Populated Western Rangelands? *Rangelands* vol. 17, pp. 148–50.

Shi, Y-B. (2000). *Amphibian Metamorphosis: From Morphology to Molecular Biology*. New York: Wiley-Liss.

Shorrocks, Bryan (2009). The Behaviour of Reticulated Giraffe in the Laikipia District of Kenya, *Giraffa* vol. 3, pp. 22–4.

Shoshani, Jeheskel, Ed. (1992). *Elephants*. Emmaus, PA: Rodale Press.

Shoshani, Jeheskel (1997). It's a Nose! It's a Hand! It's an Elephant's Trunk! *Natural History* November, pp. 36–44.

Shoshani, Jeheskel et al. (2006). Elephant Brain Part 1: Gross Morphology, Functions, Comparative Anatomy, and Evolution, *Brain Research Bulletin* vol. 70, pp. 124–57.

Shriver, Adam (2009). Knocking Out Pain in Livestock: Can Technology Succeed Where Morality has Stalled? *Neuroethics* vol. 2, pp. 115–22. https://doi.org/10.1007/s12152-009-9048-6.

Shubin, N. H. and F. A. Jenkins (1995). An Early Jurassic Jumping Frog, *Nature* vol. 377, pp. 49-52.

Siebert, Charles (2006). An Elephant Crackup? *New York Times Magazine* vol. 8, pp. 42–8.

Sigurdsen, T. and J. R. Bolt (2010). The Lower Permian Amphibamid *Doleserpeton* (Temnospondyli: Dissorophoidea), the Interrelationships of Amphibamids, and the Origin of Modern Amphibians, *Journal of Vertebrate Paleontology* vol. 30, pp. 1360–77.

Sikes, Sylvia (1971). *The Natural History of the African Elephant.* New York: American Elsevier Publishing Company, Inc.

Simmons, Robert and Lue Scheepers (1996). Winning by a Neck: Sexual Selection in the Evolution of the Giraffe, *The American Naturalist* vol. 148, pp. 771–86.

Simmons, Robert and R. Altwegg (2010). Necks-for-sex or Competing Browsers? A Critique of Ideas on the Evolution of Giraffe, *Journal of Zoology* vol. 282, pp. 6–12.

Sinclair, A. R. E. et al. (2003). Patterns of Predation in a Diverse Predator-prey System, *Nature* vol. 425, pp. 288–90. doi: 10.1038/nature01934.

Sisson, Septimus and James Grossman (1953). *The Anatomy of the Domestic Animals* (4th edition). Philadelphia, PA: W. B. Saunders Company.

Skinner, Alanson (1915). Societies of the Iowa, Kansa, and Ponca Indians, *Anthropological Papers of the American Museum of Natural History* vol. 11, pp. 679–801.

Slijper, E. J. (1946). Comparative Biologic-Anatomical Investigations on the Vertebral Column and Spinal Musculature of Mammals, *Kon. Ned. Akad. Wet., Verh. (Tweed Sectie)* DI. XLII, No. 5, pp. 1–128.

Sloan, Douglas (2015). *The Redemption of Animals.* Great Barrington, MA: Lindisfarne Books.

Slotow, Rob et al. (2000). Older Bull Elephants Control Young Males, *Nature* vol. 408, pp. 425–6.

Smuts, G. L. (1975). Pre- and Postnatal Growth Phenomena of Burchell's Zebra (*Equus burchelli antiquorum*), *Koedoe* vol. 18(1), pp. 69–102. doi: 10.4102/koedoe. v18i1.915.

Solounias, Nikos (1999). The Remarkable Anatomy of the Giraffe's Neck, *Journal of Zoology, London* vol. 247, pp. 257–68.

Solounia, Nikos et al. (1995). The Oldest Bovid from the Siwaliks, Pakistan. *Journal of Vertebrate Paleontology* vol. 15, pp. 806–14.

Soltis, Joseph (2010). Vocal Communication in African Elephants (*Loxodonta Africana*), *Zoo Biology* vol. 29, pp. 192–209.

Sordillo, Lorraine and Stacey Aitkin (2009). Impact of Oxidative Stress on the Health and Immune Function of Dairy Cattle, *Veterinary Immunology and Immunopathology* vol. 128, pp. 104–9.

Spengler Neff, A. et al. (2016). Why Cows Have Horns, *FiBL and Demeter*, December, pp. 1–16.

Spinage, C. A. (1968). *The Book of the Giraffe*. London: Collins.

Stafford, Kevin J. and David J. Mellor (2011). Addressing the Pain Associated with Disbudding and Dehorning in Cattle, *Applied Animal Behaviour Science* vol. 135, pp. 226–31.

Steiner, Rudolf (1996). *The Science of Knowing*. Spring Valley, NY: Mercury Press.

——— (2011). *The Philosophy of Freedom*. London: Rudolf Steiner Press. (This is the translation by Michael Wilson, of which earlier editions exist; there are also other translations of this book, which was originally published in 1894 in German under the title: *Die Philosophie der Freiheit*.)

Stevens, Jane (1993). Familiar Strangers, *International Wildlife* vol. 23, pp. 4–11.

Stewart, Omer C. (2009). *Forgotten Fires: Native Americans and the Transient Wilderness*. Norman, OK: University of Oklahoma Press.

Storz, B. L. et al. (2011). Reassessment of the Environmental Model of Developmental Polyphenism in Spadefoot Toad Tadpoles, *Oecologia* vol. 165, pp. 55–66. doi: 10.1007/s00442-010-1766-2.

Straus, Erwin (1963). *The Primary World of the Senses*. London: The Free Press of Glencoe.

Straus, W. L. (1949). The Riddle of Man's Ancestry, *Quarterly Review of Biology* vol. 24, pp. 256–61.

Suchantke, Andreas (2001). *Eco-Geography*. Great Barrington, MA: Lindisfarne Books.

——— (2009). *Metamorphosis: Evolution in Action*. Ghent, NY: Adonis Press.

Sukumar, Raman (1995). *Elephant Days and Nights*. Delhi: Oxford University Press.

Sunquist, Fiona (1986). Secret Energy of the Sloth. *International Wildlife* vol. 16, pp. 6–10.

Sunquist, M. E. and G. G. Montgomery (1973). Activity Patterns and Rates of Movement of Two-toed and Three-toed Sloths (*Choloepus hoffmanni* and *Bradypus infuscatus*), *Journal of Mammalogy* vol. 54, pp. 946–54.

Sutherland, Mhairi A. and Cassandra Tucker (2011). The Long and Short of It: A Review of Tail Docking in Farm Animals, *Applied Animal Behaviour Science* vol. 135, pp. 179–91.

Swenson, Melvin J., Ed. (1990). *Dukes' Physiology of Domestic Animals*, Tenth Edition. Ithaca, NY: Comstock Publishing Company and Cornell University Press.

Talbott, Stephen L. (2015). From Genes to Evolution: The Story You Haven't Heard. Available online: http://RediscoveringLife.org/ar/2015/genes_29.htm.

Tank, W. (1984). *Tieranatomie für Künstler*. Ravensburg: Otto Maier Verlag.

Taube, Erica et.al. (2001). Reproductive Biology and Postnatal Development in Sloths, *Bradypus* and *Choloepus*: Review with Original Data from the Field (French Guiana) and from Captivity, *Mammal Review* vol. 31, pp. 173–88.

Teichmann, F. (1989). Die Entstehung des Entwicklungsgedankens in der Goethezeit. In: Wolfgang Arnold, Ed. (1989). *Entwicklung: Interdisziplinäre Aspekte zur Evolutionsfrage*. Stuttgart: Urachhaus, pp. 11–26.

Tirler, Hermann (1966). *A Sloth in the Family*. London: The Harvill Press.

Trager, Matthew D. et al. (2004). Concurrent Effects of Fire Regime, Grazing and Bison Wallowing on Tallgrass Prairie Vegetation, *The American Midland Naturalist* vol. 152, pp. 237–47.

Turner, J. Scott (2017). *Purpose and Desire*. New York: HarperCollins Publishers.

Urbani, B. and C. Bosque (2007). Feeding Ecology and Postural Behaviour of the Three-toed Sloth (*Bradypus variegatus flaccidus*) in Northern Venezuela, *Mammalian Biology* vol. 72, pp. 321–9.

U.S. Department of Agriculture (2009). Reproduction Practices on U.S. Dairy Operations, 2007, *Animal and Plant Health Inspection Service*, February, Info Sheet.

———— (2014a). 2012 Census of Agriculture, *United States Summary and State Data*. Vol. 1, Geographic Area Series, Part 51 (AC-12-A-51).

———— (2014b). Dairy 2014: Health and Management Practices on U.S. Dairy Operations, 2014. http://www.aphis.usda.gov/animal_health/nahms/dairy/downloads/dairy14/Dairy14_dr_PartIII.pdf.

———— (2016). *Practices Dairy 2014: Dairy Cattle Management in the United States*, 2014. Report 1.

U.S. Food and Drug Administration. *2016 Summary Report on Antimicrobials Sold or Distributed for Use in Food-Producing Animals*, December, pp. 31–41.

Van Boeckel, Thomas P. et al. (2015). Global Trends in Antimicrobial Use in Food Animals, *Proceedings of the National Academy of Sciences* vol. 112, pp. 5649–54.

Van Boeckel, Thomas P. et al. (2019). Global Trends in Antimicrobial Resistance in Animals in Low and Middle-Income Countries. *Science* 365eaaaw1944. doi: 10.1126/science.aaw1944.

van Critters, Robert L. et al. (1966). Blood Pressure Responses of Wild Giraffes, *Science* vol. 152, pp. 384–6.

Van der Merwe, N. J. et al. (1995). The Skull and Mandible of the African Elephant (*Loxodonta africana*), *Onderstepoort Journal of Veterinary Research* vol. 62, pp. 245–60.

VanderWaal, Kimberly et al. (2014). Multilevel Social Organization and Space Use in Reticulated Giraffe (*Giraffa camelopardalis*), *Behavioral Ecology* vol. 25, pp. 17–26.

van Schalkwyk, O. L., et al. (2004). A Comparison of the Bone Density and Morphology of Giraffe (*Giraffa camelopardalis*) and Buffalo (*Syncerus caffer*) Skeletons. *Journal of Zoology, London* vol. 264, pp. 307–315.

Varela-Lasheras, Irma et al. (2011). Breaking Evolutionary and Pleiotropic Constraints in Mammals: On Sloths, Manatees and Homeotic Mutations, *EvoDevo* vol. 2, pp. 11ff. http://www.evodevojournal.com/content/2/1/11.

Verhulst, Jos (2003). *Developmental Dynamics in Humans and other Primates.* Ghent, NY: Adonis Press.

von Muggenthaler, E., et al. (1999). Infrasound and Low Frequency Vocalizations from the Giraffe. *Proceedings of the Riverbank Consortium.* Available online at http://www.animalvoice.com/Giraffe.htm.

Waage, J. K. and R. C. Best (1985). Arthropod Associates of Sloths. In: G. Gene Montgomery, Ed., *The Evolution and Ecology of Armadillos, Sloths, and Vermilinguas.* Washington D.C.: Smithsonian Institution Press, pp. 297–312.

Wackernagel, Hans (1965). Grant's Zebra (*Equus burchelli boehmi*) at Basle Zoo – A Contribution to Breeding Biology, *International Zoo Yearbook* vol. 5, pp. 38–41. doi: 10.1111/j.1748-1090.1965.tb01567.x.

Walker, James R. (1980). *Lakota Belief and Ritual.* Lincoln, NE: University of Nebraska Press.

Walsh, Denis (2015). *Organisms, Agency and Evolution.* Cambridge, UK: Cambridge University Press.

Walsh, S. W. et al. (2011). A Review of the Causes of Poor Fertility in High Milk Producing Dairy Cows, *Animal Reproduction Science* vol. 123, pp. 127–38.

Warren, James V. (1974). The Physiology of the Giraffe. *Scientific American* vol. 231(5), pp. 96–105.

Wathes, D. C. (2012). Mechanisms Linking Metabolic Status and Disease with Reproductive Outcome in the Dairy Cow, *Reproduction in Domestic Animals* vol. 47(4), pp. 304–12.

Webster, Gerry and Brian Goodwin (1996). *Form and Transformation: Generative and Relational Principles in Biology.* Cambridge, UK: Cambridge University Press.

Weiss, Paul (1973). *The Science of Life.* Mt. Kisco, New York: Futura Publishing.

White, T. D. et al. (2015). Neither Chimpanzee nor Human, *Ardipithecus* Reveals the Surprising Ancestry of Both, *PNAS* vol. 112, pp 4877–84. doi: 10.1073/pnas.1403659111.

Whitehead, Alfred North (1967). *Science and the Modern World.* New York: The Free Press.

Whitham, Thomas G., et al. (2003). Community and Ecosystem Genetics: A Consequence of the Extended Phenotype. *Ecology* vol. 84, pp. 559–573.

Whitney, George G. (1996). *From Coastal Wilderness to Fruited Plain: A History of Environmental Change in Temperate North America from 1500 to the Present.* Cambridge, UK: Cambridge University Press.

Wilkinson, David and Graeme Ruxton (2012). Understanding Selection for Long Necks in Different Taxa, *Biological Reviews* vol. 87, pp. 616–30.

Williams, David M. and Malte C. Ebach (2009). What, Exactly, is Cladistics? Rewriting the History of Systematics and Biogeography, *Acta Biotheoretica* vol. 57, pp. 249–68. doi:10.1007/s10441-008-9058-5.

Williams, Edgar (2016). Giraffe Stature and Neck Elongation: Vigilance as an Evolutionary Mechanism, *Biology* vol. 5, pp. 35–40.

Williams, J. H. (1950). *Elephant Bill*. Garden City, NY: Doubleday & Company.

Witschi, E. (1956). *Development of Vertebrates*. Philadelphia: W. B. Saunders Company.

Woolnough, A. P., and J. T. du Toit (2001). Vertical Zonation of Browse Quality in Tree Canopies Exposed to a Size-Structured Guild of African Browsing Ungulates, *Oecologia* vol. 129, pp. 595–590.

Young, J. Z. (1973). *The Life of Vertebrates* (second edition). Oxford: Clarendon Press.

Young, Truman, and Bell Okello (1998). Relaxation of an Induced Defense after Exclusion of Herbivores: Spines on *Acacia drepanolobium*, *Oecologia* vol. 115, pp. 508–513.

Young, Truman, and Lynne Isbell (1991). Sex Differences in Giraffe Feeding Ecology: Energetic and Social Constraints, *Ethology* vol. 87, pp. 79–89.

Zedeño, María N. (2008). Bundled Worlds: The Roles and Interactions of Complex Objects from the North American Plains, *Journal of Archaeological Method and Theory* vol. 15, pp. 362–78.

Zeder, Melinda A. (2011). The Origins of Agriculture in the Near East, *Current Anthropology* vol. 52, pp. S221–35.

——— (2012). *Biodiversity in Agriculture: Domestication, Evolution, and Sustainability*. Cambridge, MA: Cambridge University Press.

Zisweiler, V. (1976). *Spezielle Zoologie Wirbeltiere, Bd. I: Anamnia*. Stuttgart: Thieme Verlag.

Index

An index contains pointers to specific content in a book. When I was making the entries for this index, especially those related to the individual animals, it was clear that I was dissecting the animals into discrete parts (head, stomach, digestion, movement, etc.). And yet, the whole thrust of the book is to see these "parts" as members of a whole; that is, their true meaning and relevance is in relation to the whole. Nonetheless, I decided not to do away with these decontextualized lists, since they can be helpful for a reader to find specific facts or concepts. That said, also be forewarned that you won't really be learning anything significant if you don't study how the parts weave together and are expressive of each other and of the animal as a whole.